T0144491

Blockchain Technology and Applications

RIVER PUBLISHERS SERIES IN SECURITY AND DIGITAL FORENSICS

Indexing: All books published in this series are submitted to the Web of Science Book Citation Index (BkCI), to SCOPUS, to CrossRef and to Google Scholar for evaluation and indexing.

The “River Publishers Series in Security and Digital Forensics” is a series of comprehensive academic and professional books which focus on the theory and applications of Cyber Security, including Data Security, Mobile and Network Security, Cryptography and Digital Forensics. Topics in Prevention and Threat Management are also included in the scope of the book series, as are general business Standards in this domain.

Books published in the series include research monographs, edited volumes, handbooks and textbooks. The books provide professionals, researchers, educators, and advanced students in the field with an invaluable insight into the latest research and developments.

Topics covered in the series include, but are by no means restricted to the following:

- Cyber Security
- Digital Forensics
- Cryptography
- Blockchain
- IoT Security
- Network Security
- Mobile Security
- Data and App Security
- Threat Management
- Standardization
- Privacy
- Software Security
- Hardware Security

For a list of other books in this series, visit www.riverpublishers.com

Blockchain Technology and Applications

Ahmed Banafa

Professor of Engineering at San Jose State University (USA)

and

Instructor of Continuing Studies at Stanford University (USA)

Published, sold and distributed by:
River Publishers
Alsbjergvej 10
9260 Gistrup - Denmark
www.riverpublishers.com

ISBN: 9788770221061
e-ISBN: 9788770221054

"If you can't explain it simply,
you don't understand it well enough."
Albert Einstein

Contents

Special Topic in Blockchain

Content is king. We are the product offering at Facebook, Twitter, LinkedIn, Instagram, Yelp, YouTube and so on. When serving-up our conversations, photos, memories, and perspectives we are each providing the reason to visit these platforms. Yet, very little time is taken to consider data ownership, privacy, or our individual economic incentives. Instead, our valuable data is taken, our attention heavily shifted to the screen and we're targeted with adds ever so precisely. Blockchain provides alternative solutions to approaching the world as we know it today. And, like most products and platforms, what its creators intended may not be what it becomes. It's for the adopters to decide what to do with it. Whether its used for enterprise consortiums to improve communication in supply chains, decoupling currency from country, or finding a simpler way to transfer rewards and tokens amongst platforms – blockchain is still looking for its unicorn. We are getting closer, though. As PayPal now offers the ability to buy and sell crypto, Reddit is testing rewards on Ethereum, or JP Morgan directly servicing crypto customers Coinbase and Gemini. As you dive into the world of blockchain and crypto currency, my friend and colleague Prof. Ahmed Banafa will be the most excellent tour guide – you're going to enjoy the read!

Elizabeth "Liz" Kukka
Executive Director,
Ethereum Classic Labs | Principal Investor,
Digital Finance Group
Advisor, Matrix Exchange

Abbreviations

AI	Artificial Intelligence
API	Application Programming Interface
ATTP	Advanced Track and Trace for Pharmaceuticals
BCH	Bitcoin Cash
BFT	Byzantine Fault Tolerance
BTC	Bitcoin
CDC	Center of Disease Control
COVID19	Corona Virus Disease 2019
CSS	Cascading Style Sheets
DAO	Decentralized Autonomous Organization
DApps	Decentralized Applications
DDoS	Distributed Denial of Service
DLT	Distributed Ledger Technology
DoS	Denial of Service
DPoS	Delegated Proof of Stake
DX	Digital Transformation
ECDSA	Elliptic Curve Digital Signature Algorithm
EM	Electromagnetic
ETC	Ethereum Classic
ETH	Ethereum
EU	European Union
FDA	Food and Drug Administration
HTML	Hypertext Markup Language
IBAC	Internet of Things, Blockchain, Artificial Intelligence, Cybersecurity
IBM	International Business Machines
IDC	International Data Corporation
IEEE	Institute of Electrical and Electronics Engineering
IoT	Internet of Things
IPFS	InterPlanetary File System
IT	Information Technology
LPoS	Leased Proof of Stake
LTE	Long-Term Evolution
M2M	Machine to Machine
MIT	Massachusetts Institute of Technology
NASDAQ	National Association of Securities Dealers Automated Quotations exchange
P2P	Peer-To-Peer
PoA	Proof of Authority
PoA	Proof of Assignment
PoA	Proof of Activity
PoB	Proof of Burn
PoC	Proof of Capacity or Proof-of Concept
PoET	Proof of Elapsed Time
PoI	Proof of Importance
PoS	Proof of Stake
PoV	Proof of View
PoW	Proof of Work
Qubit	Quantum Bit
SSD	Solid State Drive
UI	User Interface
UX	User Experience
WHO	World Health Organization
ZKP	Zero-Knowledge Proof or Protocol

List of Figures and Tables

Preface

Blockchain is an emerging technology that can radically improve transaction security at banking, supply chain, and other transaction networks. It's estimated that Blockchain will generate $3.1 trillion in new business value by 2030. Essentially, it provides the basis for a dynamic distributed ledger that can be applied to save time when recording transactions between parties, remove costs associated with intermediaries and reduce risks of fraud and tampering. This book explores the fundamentals and applications of Blockchain technology. Readers will learn about the decentralized peer-to-peer network, distributed ledger, and the trust model that defines Blockchain technology. They will also be introduced to the basic components of Blockchain (transaction, block, block header, and the chain), its operations (hashing, verification, validation, and consensus model), underlying algorithms, and essentials of trust (hard fork and soft fork). Private and public Blockchain networks similar to Bitcoin and Ethereum will be introduced, as will concepts of Smart Contracts, Proof of Work and Proof of Stack, and cryptocurrency including Facebook's Libra will be elucidated. Also, the book will address the relationship between Blockchain technology, the Internet of Things (IoT), Artificial Intelligence (AI), Cybersecurity, Digital Transformation, and Quantum Computing.

Readers will understand the inner workings and applications of this disruptive technology and its potential impact on all aspects of the business world and society. A look at the future trends of Blockchain Technology will be presented in the book.

The book can be divided into 3 parts:

Blockchain Technology

Chapter 1 Introduction to Blockchain
Chapter 2 Consensus Protocols
Chapter 3 Key Blockchain Use Cases
Chapter 4 Important Topics in Blockchain
Chapter 5 Decentralized Applications – DApps

Blockchain Applications

Chapter 6: Using Blockchain to Secure IoT
Chapter 7: IoT and Blockchain: Challenges and Risks
Chapter 8: IoT, AI and Blockchain: Catalysts for Digital Transformation
Chapter 9: Myths about Blockchain Technology
Chapter 10: Cybersecurity & Blockchain
Chapter 11: Blockchain and AI: A Perfect Match?
Chapter 12: Quantum Computing and Blockchain: Facts and Myths
Chapter 13: Cryptocurrency: To Libra or not To Libra
Chapter 14: Future Trends of Blockchain

Special Topic in Blockchain

Chapter 15: Blockchain Technology and COVID-19

Audience

This is book is for everyone who would like to have a good understanding of Blockchain Technology and its applications and its relationship with business operations including C-Suite executives, IT managers, marketing & salespeople, lawyers, product & project managers, business specialists, students. It is not for programmers who are looking for codes or exercises on the different platforms of Blockchain.

Acknowledgment

I am grateful for all the support I received from my mother, my wife, and my children while writing this book.

I dedicate this book to my late father.

Part 1

Blockchain Technology

1

Introduction to Blockchain

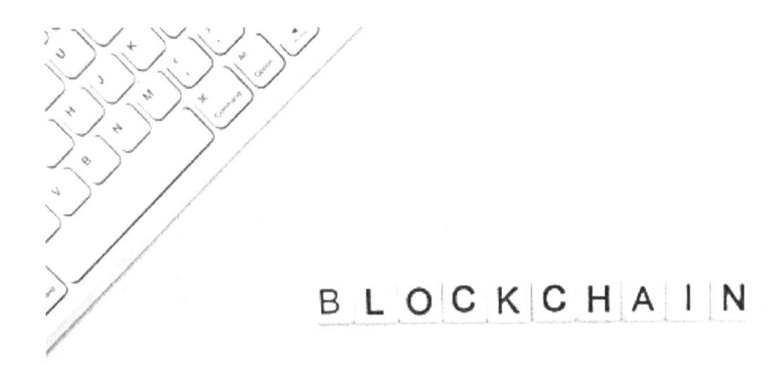

Blockchain Technology is one of the four hot technologies shaping the future of the tech world in the coming decades, these four technologies (IBAC) are: Internet of Things (IoT), Blockchain, Artificial Intelligence (AI), and Cybersecurity (Figure 1.1). All four technologies are interconnected and impact each other in many ways. As Figure 1.2 shows that you can explain each technology with an analogy to human acts: IoT: Feels, Blockchain: Remembers, AI: Thinks, and Cybersecurity: Protects.

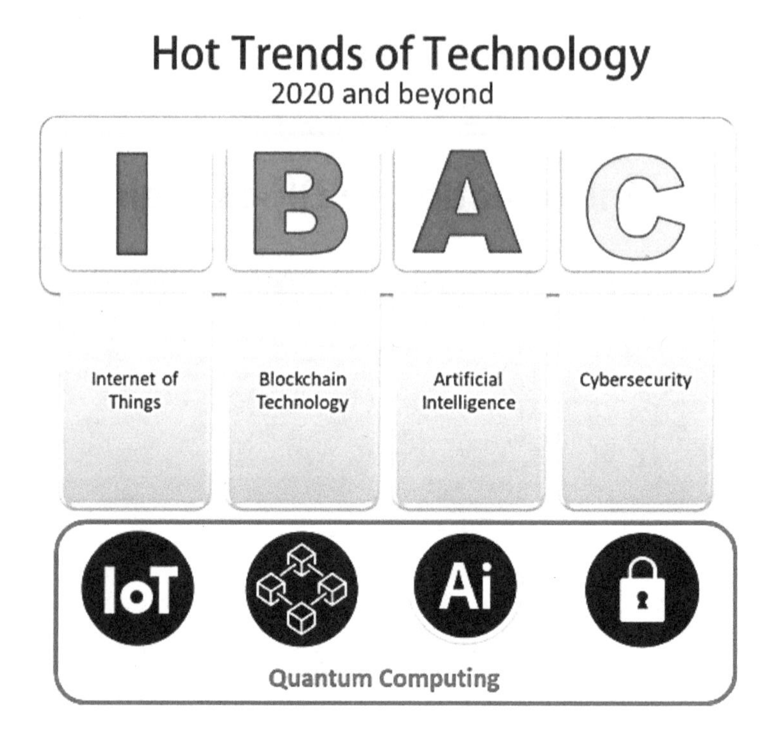

Figure 1.1: Simplified form of IBAC

Recently, "Quantum Computing" presented itself as a new player impacting IBAC in many ways, for example, Quantum Computing will make IoT faster in processing data and extracting insights, Quantum Computing will force Blockchain to invent new encryption techniques and will make processing data faster solving one of the main issues of Blockchain Technology, in the case of AI Quantum Computing will make analysis extremely faster which will, in turn, makes decisions done real-time in many cases not possible with current computing tools, in Cybersecurity, Quantum Computing will help in detection and prevention of cyber-attacks and open the doors for new Quantum Encryptions algorithms which will make it very hard for hackers to access systems and data.

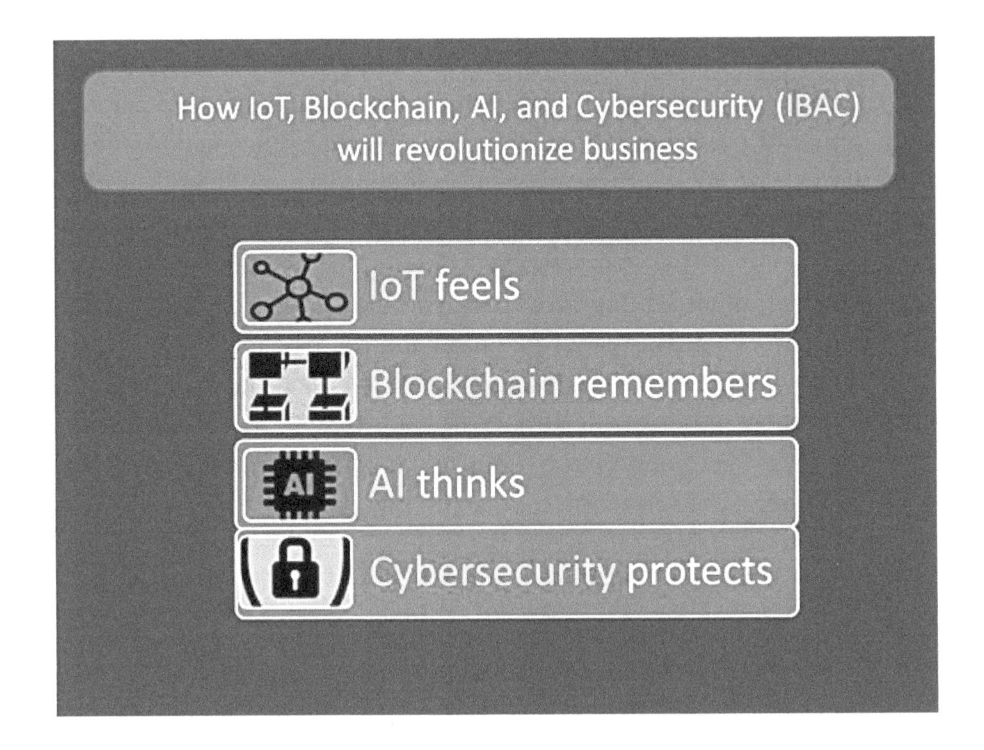

Figure 1.2: Hot Trends of Technology in 2020 and Beyond

1.1 What is Blockchain?

Blockchain is simply a software to start with the classical definition of Blockchain is "a distributed database existing on multiple computers at the same time. It is constantly growing as new sets of recordings, or 'blocks', are added to it. Each block contains a timestamp and a link to the previous block, so they actually form a chain", but the best definition of Blockchain according to MIT is: *Cryptography +Human Logic*.

If the internet is all about providing *connectivity*, Blockchain is all about enabling *trust*. For example, imagine there are 30 people in a classroom or an office building, with one main door and a security guard holding a list of authorized students/employees who can get into the building, you will show your card to him/her to check the list and if you are on the list you are in. This is the current centralized system. With the use of Blockchain, each one of the 30 people will have a list with pictures of people who are authorized to be in the room so if somebody came in, and that person was not on the list, they would start talking to each other, asking "Hey, can you please check if this person belongs here?" That is a synchronization and referred to as gossip protocol within the Blockchain. Human logic is the list you have, and the motion of everybody starting to talk to each other. On the top of the current system using encryption (user name and password), we added the human logic, consensus protocols and algorithms.

Figure 1.3: Best Definition of Blockchain

1.2 The Five Components of a Blockchain

1. Cryptography
2. P2P Network
3. Consensus Mechanism
4. Ledger
5. Validity Rules

All listed in Figure 1.4.

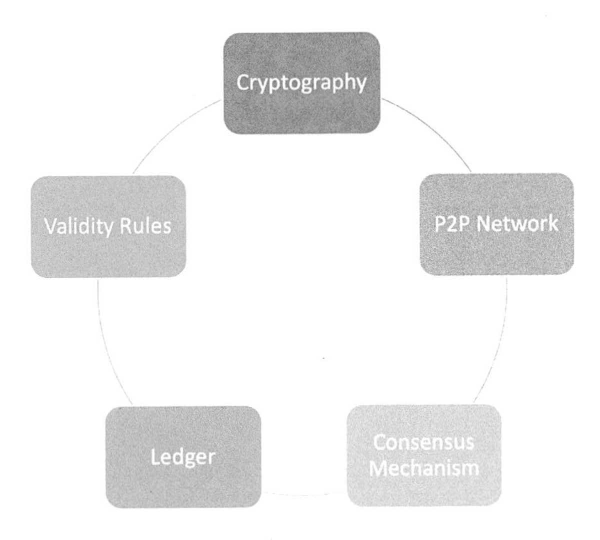

Figure 1.4: Five Components of a Blockchain

1.3 Blockchain Programming Languages

Any of the following programming languages can be used to create Blockchain platforms:

- C++ (Bitcoin)
- Python
- JavaScript
- Solidity (Smart Contract)
- Java
- Go

Figure 1.5 shows example of a “block” programming.

```
class Block {
        constructor(timestamp, transactions,
        previousHash = '') {
        this.previousHash = previousHash;
        this.timestamp = timestamp;
        this.transactions = transactions;
        this.hash = this.calculateHash();
        this.nonce = 0;
        }
}
```

Figure 1.5: Example of a "block" programming

1.4 Mechanism of Blockchain Technology

First block called Genesis Block, created by the miner or validator based on consensuses protocol, each block have five elements (Index, Time-Stamp, Previous Hash, Hash, and Data), a Blockchain is initialized with the genesis block which is the foundation of the trading system and the prototype for the other blocks in the Blockchain. When you change any of these data's you will change the whole block and the following blocks will see that something has changed, in addition to the other nodes with copies of the blocks and the altered node will be rejected, all nodes sync using a gossip protocol, Figure 1.6 shows this type of mechanism.

A gossip protocol is a procedure or process of computer peer-to-peer communication that is based on the way epidemics spread. Some distributed systems including Blockchain use peer-to-peer gossip to ensure that data is disseminated to all members of a group. Some ad-hoc networks have no central registry and the only way to spread common data is to rely on each member to pass it along to their neighbors. [1]

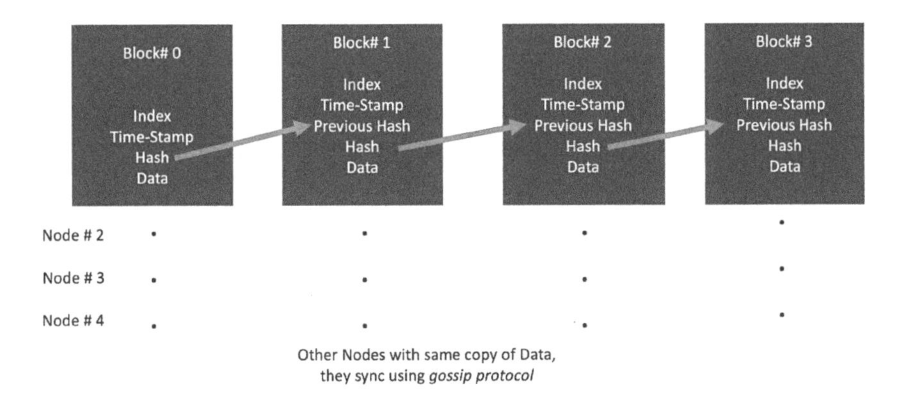

Figure 1.6: Example of Blocks of Blockchain in one node

1.5 Blockchain vs. Traditional Database

There are many differences between Blockchain and Traditional Database and Table 1.1 summarizes them:

Blockchain vs. Traditional Database		
Characteristics	**Blockchain**	**Database**
Authority	Decentralized	Centralized and controlled by the admin
Architecture	Distributed	Client-server
Data Handling	Read and Write	CRUD (Create, Read, Update, Delete)
Integrity	High	Can be altered by hackers
Transparency	High	Controlled by the admin
Cost	High	Low
Performance	Slow	Very fast

Table 1.1: Blockchain vs. Traditional Database

1.6 The Stack of Blockchain

Like any other technology, Blockchain can be defined by its stack, the following diagram explains it and it is worth mentioning to emphasize the importance of each layer as an opportunity for improvement and new business (startups), for example, UI/UX for different devices including smartphones, tablets, desktops, and laptops, in addition to the wide field of new consensus protocols for specific applications and industries, the introduction of smart contracts in the design process to avoid any surprises, and secure ways to connect the stack to the internet.

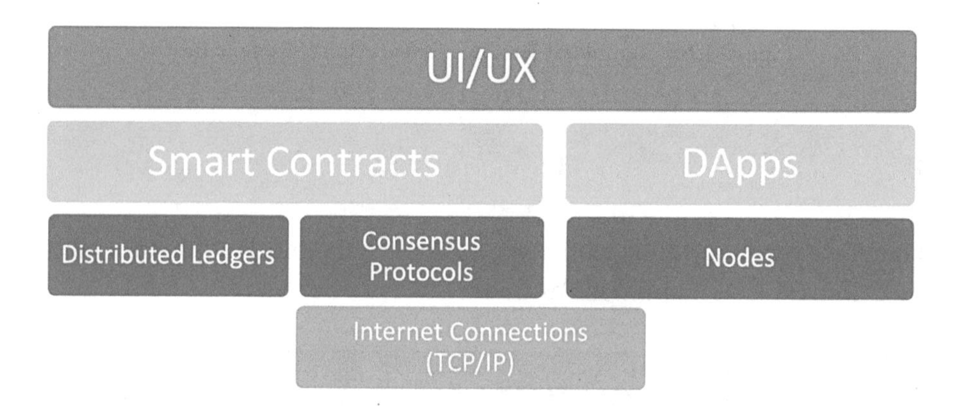

Figure 1.7: Blockchain Stack

1.7 Blockchain Tracks

To understand the future direction of Blockchain technology, we need to recognize the three tracks (Figure 1.8) of Blockchain technology:

- *Pure R&D Track*: This track is focused on understanding what it means to develop a Blockchain-based system. Ideally, working on real use-cases, the ultimate goal is investigation and learning, and not necessarily delivery of a working system.

- *Immediate Business Benefit Track*: This track covers two bases: (1) learning how to work with this promising technology and (2) delivering an actual system that can be deployed in a real business context. Many of these projects are intra-company.
- *Long-Term Transformational Potential Track*: This is the track of the visionaries, who recognize that to realize the true value of Blockchain-based networks means reinventing entire processes and industries as well as how public-sector organizations function.

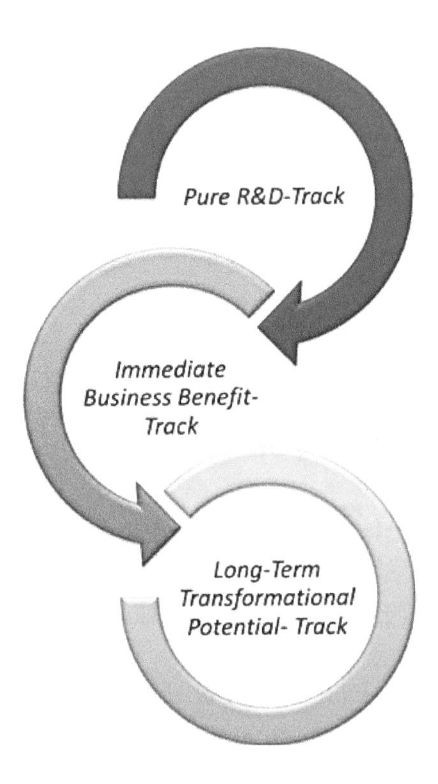

Figure 1.8: Tracks of Blockchain Technology

1.8 Challenges facing Blockchain Technology

Every new technology face challenge and Blockchain is not an exception, the following is a list of both technical and non-technical challenges (Figure 1.9):

Technical Challenges

- Scalability
- Processing Time
- Processing Power
- 51% Attack
- Double Spending
- Bad Smart Contracts
- Storage
- First Mile and Last Mile problem (Data before and after going through the Blockchain)

Non-Technical Challenges

- Regulations
- Public perception (Blockchain is Bitcoin)
- Lack of skilled staff

1.9 Types of Blockchain Networks

There are three types of Blockchain Networks (Figure 1.10) :

- **Public:** a public Blockchain is the one where everyone can see all the transactions, anyone can expect their transaction to appear on the ledger and finally anyone can participate in the consensus process.

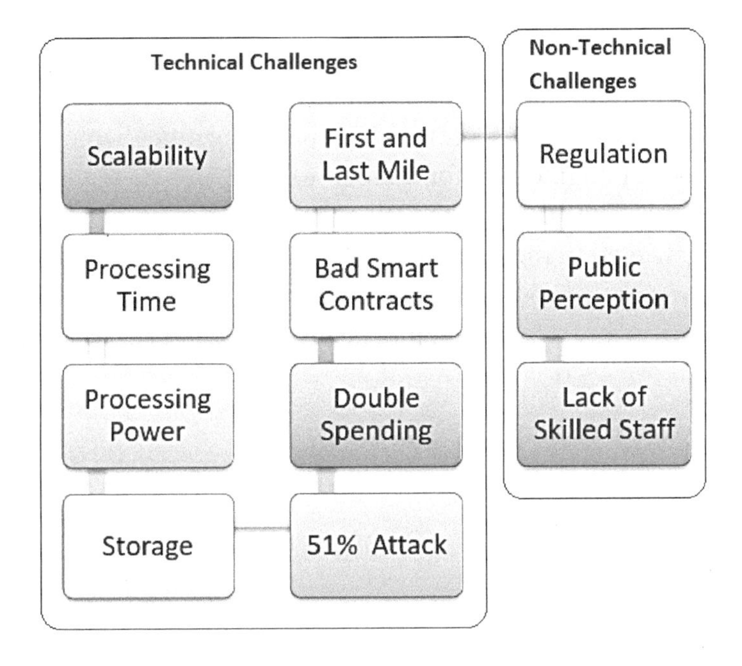

Figure 1.9: Challenges Facing Blockchain

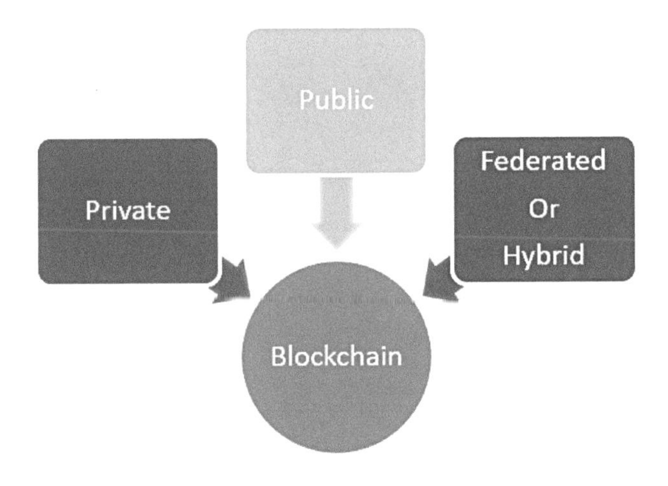

Figure 1.10: Types of Blockchain Networks

- **Federated/Hybrid:** federated/hybrid Blockchain does not allow everyone to participate in the consensus process. Indeed, only a limited number of nodes are given permission to do so. For instance, in a group of 20 pharmaceutical companies, we could imagine that for a block to be valid, 15 of them have to agree. The access to the Blockchain, however, can be public or restricted to the participants.
- **Private:** private Blockchains are generally used inside a company. Only specific members are allowed to access it and carry out transactions.

2

Consensus Protocols

A consensus protocol may be defined as the mechanism through which a Blockchain network reaches consensus. Blockchains are built as distributed systems and, since they do not rely on a central authority, the distributed nodes need to agree on the validity of transactions.

This is where consensus protocols come into play. They assure that the protocol rules are being followed and guarantee that all transactions occur in a trustless way.

2.1 Types of Consensus Algorithms [2]

Below is a list of consensus algorithms, there are many other algorithms or protocols beside the ones listed here depending on the specific application and use case of Blockchain. Proof of X, where "X" can be any requirements, is the most exciting field of research for students and researches where creativity plays a major part in creating new consensus algorithms.

2.1.1 Proof of Work

Most cryptocurrencies including Bitcoin run on "proof of work". Proof of work as a process has the following steps to it:

- The miners solve cryptographic puzzles to "mine" a block to add to the Blockchain.
- This process requires an immense amount of energy and computational usage.
- The puzzles have been designed in a way that makes it hard and taxing on the system.
- When a miner solves the puzzle, they present their block to the network for verification.

Mining serves as two purposes:

1. To verify the legitimacy of a transaction, or avoiding the so-called *double-spending*;
2. To create *new digital currencies* by rewarding miners for performing the previous task.

The miner must find a result starting with a number of zeroes.

The greater the number of zeroes, the more difficult it is for the miner to find the result and the more it will have to try his luck before finding it.

Yet the number of zeroes (and therefore the difficulty) is adjusted to the number of miners on the network (and their computer capacity or hashing power) to be sure that it will take an average of 10 minutes to find the solution. Once it has found this figure, the other members of the network can instantly check the solution.

Since the miner may not find the Input data ("input") from the result ("output"), he/she is going to try his/her luck until he located the Input data enabling him/her to obtain the output data corresponding to the objective of difficulty required, which is the number starting with a number of zeroes sufficient to be validated by the Protocol Bitcoin and thus be added to the Blockchain.

Figure 2.1 is an example of the mathematical puzzle, the goal is to have 3 leading zeroes:

This city is amazing1= 0ndldeouklewnlf88980378008mmkjj...

This city is amazing2= 0ljljfdijirejopnjojnojre9980089knlkd9...

This city is amazing3= 0uuuiiasmlmnp122339u0unnklmnkj...

.

.

.

This city is amazing409876345921 = 000jukutyghi7j6544ghjjj239...

Figure 2.1: Example of Mathematical puzzle in Proof of Work (PoW) Protocol

2.1.2 Proof of Stake

Proof of stake will make the entire mining process virtual and replace miners with validators.

This is how the process will work:

1. The validators will have to *lock up* some of their coins as stake.
2. After that, they will start validating the blocks. Meaning, when they discover a block that they think can be added to the chain, they will validate it by placing a bet on it.
3. If the block gets appended, then the validators will get a reward proportionate to their bets.

Proof of stake is a different way to validate transactions to achieve the distributed consensus. It is still an algorithm, and the purpose is the same as the proof of work, but the process to reach the goal is quite different.

Unlike the proof of work, where the algorithm rewards miners who solve mathematical problems with the goal of validating transactions and creating new blocks, with the proof of stake, the creator of a new block is chosen in a deterministic way, depending on its wealth, also defined as stake. No block rewards.

Also, all the digital currencies are previously created in the beginning, and their number never changes. This means that in the PoS system there is no block reward, so, the miners take the transaction fees.

2.1.3 Delegated Proof of Stake (DPoS)

DPoS is similar to PoS in regard to staking but has a different and more democratic system that is said to be fair. Like PoS, token holders stake their tokens in this consensus protocol.

Instead of the probabilistic algorithm in PoS, token holders within a DPoS network are able to cast votes proportional to their stake to appoint delegates to serve on a panel of witnesses—these witnesses secure the Blockchain network. In DPoS, delegates do not need to have a large stake, but they must compete to gain the most votes from users.

It provides better scalability compared to PoW and PoS as there are fully dedicated nodes who are voted to power the Blockchain. Block producers can be voted in or out at any time, and hence the threat of tarnishing their reputation and loss of income plays a major role against bad actors.

It's clear, DPoS seem to result in a semi-centralized network, but it is traded off for scalability.

2.1.4 Proof of Authority (PoA)

PoA is known to bear many similarities to PoS and DPoS, where only a group of pre-selected authorities (called validators) secure the Blockchain and are able to produce new blocks.

New blocks on the Blockchain are created only when a supermajority is reached by the validators. The identities of all validators are public and verifiable by any third party—resulting in the validator's public identity performing the role of proof of stake.

As these validators' identities are at stake, the threat of their identity being ruined incentivizes them to act in the best interest of the network. Due to the fact that PoA's trust system is predetermined, concerns have been raised that there might be a centralized element with this consensus algorithm.

However, it can be argued that semi-centralization could actually be appropriate within private/consortium Blockchains.

2.1.5 Proof of Assignment (PoA)

Similarly, to DPoS, the proof of assignment model establishes several trusted nodes within the network, but only those nodes store the entire ledger. By allowing other network contributors to participate without ledger storage.

2.1.6 Byzantine Fault Tolerance (BFT)

BFT is the most popular permissioned (private) Blockchain platform protocol and is currently used by Hyperledger Fabric.

To understand the Byzantine Fault Tolerance algorithm, you need to understand the Byzantine generals' problem. Imagine a group of Byzantine generals and their armies have surrounded a castle and preparing to attack.

To win, they must attack simultaneously. But they know that there is at least one traitor among them. So, how do they launch a successful attack with at least one, unknown, bad actor in their midst?

The analogy is clear: In any distributed computing environment—Blockchain—there is a risk that rogue actors could wreak havoc. So, its reliance on community consensus makes Byzantine faults an especially thorny problem for Blockchain. PoW generally provides a solution: "Byzantine fault tolerance." But the drawbacks may not be worth it.

That is where the Byzantine Fault Tolerance algorithm comes into play. It is considered the first practical solution to achieving consensus that overcomes Byzantine failure.

A consensus decision is determined based on the total decisions submitted by all the generals.

It addresses the challenges without the expenditure of energy required by proof of work. But it works only on a permissioned Blockchain because there is no anonymity.

2.1.7 Leased Proof of Stake (LPoS)

Leased Proof of Stake is an advanced version of the Proof of Stake (PoS) algorithm. Generally, in the Proof of Stake algorithm, every node holds a certain amount of cryptocurrency and is suitable to add the next block into the Blockchain.

However, with Leased Proof of Stake, users are able to lease their balance to full nodes. The higher the amount that is leased, the better the chances are that the full node will be selected to produce the next block. If the node is selected, the user will receive some part of the transaction fees that are collected by the node.

2.1.8 Proof of Elapsed Time (PoET)

PoET is a consensus mechanism algorithm that is often used on the permissioned (private) Blockchain networks to determine the mining rights or the block winners on the network. Based on the basis of a fair lottery system where every single node is equally likely to be a winner, the PoET mechanism is based on spreading the chances of winning fairly across the largest possible number of network users.

The timer is different for every node. Every user in the network is assigned a random amount of time to wait, and the first user to finish waiting gets to commit the next block to the blockchain. Compare to pulling straws, but this time, the shortest stem in the stack wins the lottery.

2.1.9 Proof of Activity (PoA)

Proof of activity is one of the many Blockchain consensus algorithms used to assure that all the transactions following on the Blockchain are genuine and all users arrive at a consensus on the precise status of the public ledger.

Proof of activity is a mixed approach that conjoins the other two commonly used algorithms—proof of work (POW) and proof of stake (POS).

2.1.10 Proof of Importance (PoI)

Proof of Importance is a consensus algorithm similar to PoS. Nodes "vest" currency to participate in the creation of blocks. Unlike PoS, Proof of Importance quantifies a user's support of the network.

2.1.11 Proof of Capacity (PoC)

Proof of capacity (POC) is a consensus mechanism algorithm used in Blockchains that allows the mining devices in the network to use their usable hard drive space to decide the mining rights, instead of using the mining device's computing power (as in the proof of work algorithm) or the miner's stake in the crypto coins (as in the proof of stake algorithm).

2.1.12 Proof of Burn (PoB)

Unlike PoW, Proof of Burn (PoB) is a consensus mechanism that does not waste energy.

The real computing power is not critical to avoid manipulation. In this case, the nodes destroy or burn their tokens if they want to create the next blocks and receive a reward.

With PoB, every time a user decides to destroy a part of their tokens, they buy a part of the virtual computing power that gives them the ability to validate the blocks. The more tokens they burn, the higher the possibility of receiving the reward.

3

Key Blockchain Use Cases

Blockchain use cases fall into two fundamental categories: *record keeping*, static registries of data about highly valuable assets, and *transactions*, dynamic registries of the exchange of tradeable assets:

- Record keeping use cases include the long-term safeguarding of data on valuable physical and digital assets, keeping track of identity-related information about individuals and executable smart contracts based on pre-defined conditions.

- Transaction use cases include keeping track of data about frequently exchanged assets, near-real-time digital payments, and emerging digital assets.

Here are four ways that Blockchain is actually useful to avoid pilot to production failure.

- The first use case is for guaranteed and verified data dissemination.
- The second use case is an asset and product tracking.
- The third use case is asset transfer.
- The fourth use case is certified claims.

Next is a comprehensive list of Blockchain applications in different industries. [3]

3.1 How Blockchain Can Help Advertising

For buy-side transparency: Blockchain for auditing
For sell-side transparency: Proof of View (PoV) to fight fraud

- PoV only records views from signed-in users, since the viewer's unique ID is part of the information required for a view to be considered valid.
- Since most people are only able to watch one video at a time, the PoV will invalidate views from a user who is streaming multiple videos simultaneously.
- The PoV technology confirms that a video is actually being streamed by capturing information about the current frame at random times.
- Using smart contracts to document views and who gets paid

3.2 Verifying the Authenticity of Returned Drugs

In the Pharma industry, drugs are frequently returned to the pharmaceutical manufacturers.

While the proportion of the returned drugs is small compared to the sales (about 2–3% of sales), the per year volume is in the range of $7–10 billion.

Currently, big pharma working with tech companies to develop Pharma Blockchain Proof of Concept (POC) app for this use case:

The system generates unique identifiers for a drug package. When a manufacturer ships a package, they register the item on the Pharma POC Blockchain, with the four pieces of information generated by the ATTP; the item number (based on GS1 standard), a serial number, a batch number, and expiration date. Using this PoC tracking will be easy, efficient and fast.

3.3 Transparency and traceability of consent in Clinical Trials

Informed patient consent involves making the patient aware of each step in the Clinical Trial process including any possible risks posed by the study. Clinical trial consent for protocols and their revisions need to be transparent for patients and traceable for stakeholders.

However, in practice, the informed consent process is difficult to handle in a rigorous and satisfactory way. The FDA reports that almost 10% of the trials they monitor feature some issues related to consent collection.

These include: failure to obtain written informed consent, unapproved forms, invalid consent document, failure to re-consent to a revised protocol and missing Institutional Review Board approval to protocol changes, amongst others. Frequently also there are reported cases of document fraud such as issues of backdating consent documents.

Blockchain technology provides a mechanism for unfalsifiable time-stamping of consent forms, storing and tracking the consent in a secure and publicly verifiable way, and enabling the sharing of this information in real-time.

Additionally, smart contracts can be bound to protocol revisions, such that any change in the clinical trial protocol requires the patient consent needing renewal.

3.4 Insurance

Arguably, the greatest Blockchain application for insurance is through smart contracts. Such contracts powered by Blockchain could allow customers and insurers to manage claims in a truly transparent and secure manner, according to Deloitte.

All contracts and claims could be recorded on the Blockchain and validated by the network, which would eliminate invalid claims.

For example, the Blockchain would reject multiple claims on the same accident.

3.5 Real Estate

The average homeowner sells his or her home every 5 to 7 years, and the average person will move nearly 12 times during his or her lifetime. With such movement, Blockchain could certainly be of use in the real estate market.

It would expedite home sales by quickly verifying finances, would reduce fraud thanks to its encryption, and would offer transparency throughout the entire selling and purchasing process.

3.6 Energy

Blockchain technology could be used to execute energy supply transactions, but it could further provide the basis for metering, billing, and clearing processes, according to PWC.

Other potential applications include documenting ownership, asset management, origin guarantees, emission allowances, and renewable energy certificates.

3.7 Record Management

National, state, and local governments are responsible for maintaining individuals' records such as birth and death dates, marital status, or property transfers.

Yet managing this data can be difficult, and to this day some of these records only exist in paper form and sometimes, citizens have to physically go to their local government offices to make changes, which are time-consuming, unnecessary, and frustrating. Blockchain technology can simplify this record-keeping.

3.8 Crowdfunding

Blockchain technology, among all its benefits, can be best put to use by providing provable milestones as contingencies for giving, with smart contracts releasing funds only once milestones establish that the money is being used the way that it is said to be. By providing greater oversight into individual campaigns and reducing the amount of trust required to donate in good conscience, crowdfunding can become an even more legitimate means of funding a vast spectrum of projects and causes, Figure 3.1 shows How Blockchain Is Revolutionizing Crowdfunding. [4]

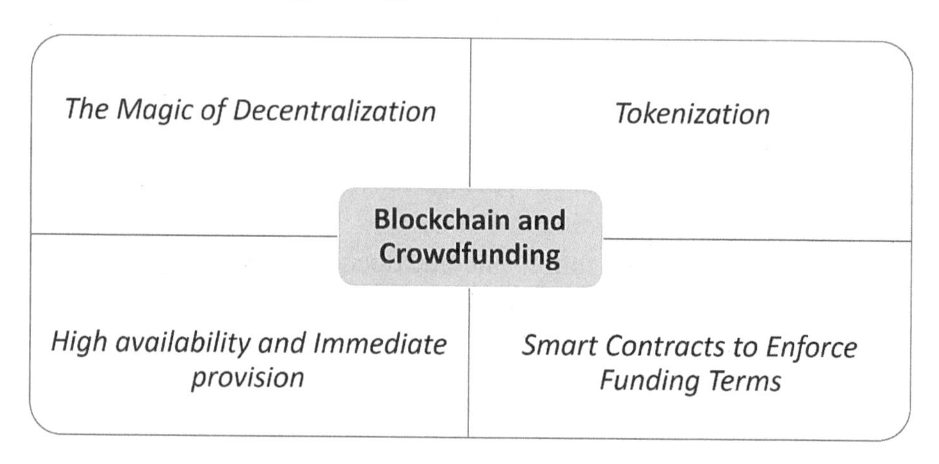

Figure 3.1: Blockchain and Crowdfunding

How Blockchain helps Crowdfunding

1. *The Magic of Decentralization:* Startups are not going to rely on any platform or combination of platforms to enable creators to raise funds. Startups no longer be beholden to the rules, regulations, and whims of the most popular crowdfunding platforms on the internet. Literally, any project has a chance of getting visibility and getting funded. It also eliminates the problem of fees. While blockchain upkeep does cost a bit of money, it will cut back drastically on transaction fees. This makes crowdfunding less expensive for creators and investors. [5]

2. *Tokenization:* Instead of using crowdfunding to enable pre-orders of upcoming tangible products, blockchain could rely on asset tokenization to provide investors with equity or some similar concept of ownership, for example Initial Coin Offering (ICO). That way, investors will see success proportional to the eventual success of the company. This could potentially

open whole new worlds of investment opportunity. Startups could save money on hiring employees by compensating them partially in fractional ownership of the business, converting it into an employee-owned enterprise. Asset tokens become their own form of currency in this model, enabling organizations to do more like hire professionals like marketers and advertisers. [5]

3. *High availability and Immediate provision:* Any project using a blockchain-based crowdfunding model can potentially get funded. Also, any person with an internet connection can contribute to those projects. Blockchain-based crowdfunders wouldn't have to worry about the "fraud" that have plagued modern-day crowdfunding projects. Instead contributors will immediately receive fractional enterprise or product ownership. [5]

4. *Smart Contracts to Enforce Funding Terms:* There are several ways in which blockchain-enabled smart contracts could provide greater accountability in crowdfunding. Primarily, these contracts would provide built-in milestones that would prevent funds from being released without provenance as to a project or campaign's legitimacy. This would prevent large sums of money from being squandered by those who are either ill-intended or not qualified to be running a crowdfunding campaign in the first place. [4]

3.9 Blockchain Technology and Supply Chain Management

Managing today's supply chains is extremely complex. For many products, the supply chain can span over hundreds of stages, multiple geographical (international) locations, a multitude of invoices and payments, have several individuals and entities involved, and extend over months of time. Due to the complexity and lack of transparency of the current supply chains, there is high interest in how Blockchains might transform the supply chain and logistics industry. [6]

This interest rose from the long list of issues with current Supply Chain Management (SCM) including [7]:

- Difficulty of Tracking
- Lack of Trust
- High Costs: procurement costs, transportation costs, inventory costs and quality costs
- Globalization Barriers

3.9.1 Blockchain and SCM

Blockchain technology and supply chain management systems were built for each other in many ways. In fact, several of the flaws of the current supply chains can be easily relieved by using Blockchain technology. Supply Chain Management (SCM) is one of the foremost industries that Blockchain can disrupts and changes for the better. [7]

With Blockchain technology properties of *decentralization*, *transparency*, and *immutability*, it is the perfect tool to save the supply chain management industry. Subsequently, Blockchain can increases the efficiency and transparency of supply chains and positively impact everything from warehousing to delivery to payment. Most importantly, Blockchain provides *consensus*—there is no dispute in the chain regarding transactions because all entities on the chain have the same version of the ledger [8].

Blockchain can have a big impact on SCM in two dimensions *Traceability*, and *Transparency (Figure 3.2)*:

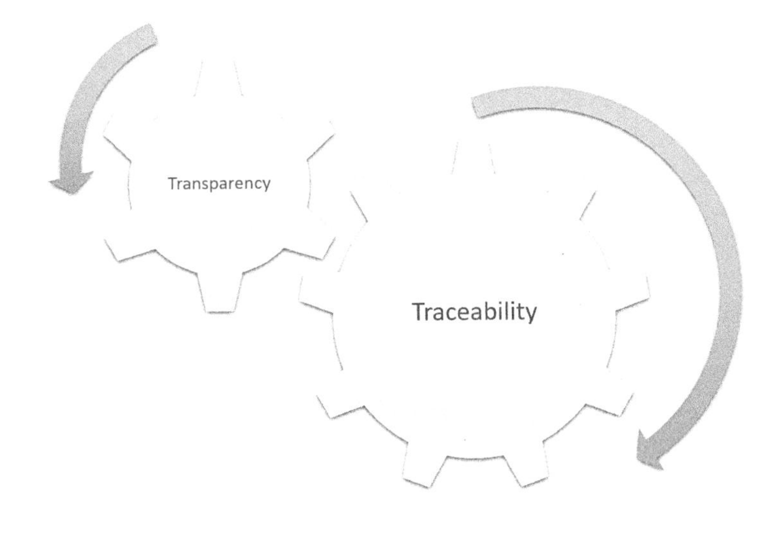

Figure 3.2: Blockchain and Supply Chain Management

Traceability Blockchain improves operational efficiency by mapping and visualizing enterprise supply chains. A growing number of consumers demand sourcing information about the products they buy. Blockchain helps organizations understand their supply chain and engage consumers with real, verifiable, and immutable data [9].

Transparency Blockchain builds trust by capturing key data points, such as certifications and claims, and then provides open access to this data publicly. Once registered on the Blockchain, its authenticity can be verified by a third-party. The information can be updated and validated in real-time. Plus, the strong security from its innate cryptography will eliminate unnecessary audits, saving copious amounts of time and money [9].

Applying Blockchain technology to SCM can results in tremendous benefits, including [9]:

- Less Time Delays
- Less Human Error
- Less Costs

4

Important Topics in Blockchain

There are many topics related to Blockchain are equally important as the technology itself, because they solve specific problems and challenges facing Blockchain, including Forks, Sharding, ZKP, and Smart Contracts.

4.1 Soft Fork vs. Hard Fork

A fork is a change to the protocol or a divergence from the previous version of the Blockchain. When a new, alternative, block is generated by a rogue miner, the system reaches consensus that this block is not valid, and this 'orphan block' is very soon abandoned by the other miners.

Forks in Blockchain are of two types: *Soft Fork* and *Hard Fork*.

4.1.1 Soft Fork

A soft fork is a software upgrade that is *backward compatible with older versions*.

This means that participants that did not upgrade to the new software will still be able to participate in validating and verifying transactions.

It is much easier to implement a soft fork as only a majority of participants need to upgrade the software. All participants, whether they have updated or not will continue to recognize new blocks and maintain compatibility with the network.

A point to take note, however, is that the functionality of a non-upgraded participant is affected. An example of a soft fork is when the new rule states that the block size will be changed from the current 1MB (1,000KB) to 800KB.

Non-upgraded participants will still continue to see that the incoming new transactions are valid. The issue is when non-upgraded miners try to mine new blocks, their blocks (and thus, efforts) will be rejected by the network.

Hence, soft forks represent a gradual upgrading mechanism as those who have yet to upgrade their software is incentivized to do so, or risk having reduced functionalities.

4.1.2 Hard Fork

Hard forks refer to a software upgrade that is not compatible with older versions.

All participants must upgrade to the new software to continue participating and validating new transactions. Those who did not upgrade would be separated from the network and cannot validate the new transactions. This separation results in a permanent divergence of the Blockchain.

As long as there is support in the minority chain – in the form of participants mining in the chain – the two chains will concurrently exist. Example: Ethereum Classic and Bitcoin Cash.

4.1.3 Ethereum Classic

Ethereum had a hard fork to reverse the effects of a hack that occurred in one of their applications (called the Decentralized Autonomous Organization or simply, DAO).

However, a minority portion of the community was philosophically opposed to changing the Blockchain at any costs, to preserve its nature of immutability.

As Ethereum's core developers and the majority of its community went ahead with the hard fork, the minority that stayed behind and did not upgrade their software continued to mine what is now known as Ethereum Classic (ETC).

It is important to note that since the majority transited to the new chain, they still retained the original ETH symbol, while the minority supporting the old chain were given the term Ethereum Classic or ETC.

4.1.4 Bitcoin Cash

Bitcoin Cash is a cryptocurrency that is a fork of Bitcoin. Bitcoin Cash is a spin-off or altcoin that was created in 2017. In 2018, Bitcoin Cash subsequently split into two cryptocurrencies: Bitcoin Cash, and Bitcoin. Bitcoin Cash is sometimes also referred to as Bcash. [10]

Bitcoin was forked to create Bitcoin Cash because the developers of Bitcoin wanted to make some important changes to Bitcoin. The developers of the Bitcoin community could not come to an agreement concerning some of the changes that they wanted to make. So, a small group of these developers forked Bitcoin to create a new version of the same code with a few modifications. [11]

The changes that make all the difference between Bitcoin Cash vs Bitcoin are these (Table 4.1):

- Bitcoin Cash has cheaper transfer fees, so making transactions in BCH will save you more money than using BTC.
- BCH has faster transfer times. So, you do not have to wait for 10 minutes it takes to verify a Bitcoin transaction!
- BCH can handle more transactions per second. This means that more people can use BCH at the same time than they can with BTC.

Table 4.1 compared both cryptocurrencies. [12]

Bitcoin	Bitcoin Cash
Standard Block Size: 1MB Max	PowerBlocks: 8 MB Max
Transactions Signatures can be discarded from the Blockchain	All transactions signatures must be validated and secured on the Blockchain
Single Centralized Development Team and Client Implementation	Multiple Independent Development Teams and client implementations

Table 4.1: Bitcoin vs. Bitcoin Cash

4.2 Zero-Knowledge Proof (ZKP)

In cryptography, a zero-knowledge proof or zero-knowledge protocol is a method by which one party (the prover) can prove to another party (the verifier) that they know a value x, without conveying any information apart from the fact that they know the value x.

Zero-knowledge proof is the ability to prove a secret without revealing what the secret is.

4.2.1 How does zero-knowledge proof work?

The best way to explain the process of zero-knowledge proofs is with a non-digital example which is, of course, far from the complexity of zero-knowledge proofs but very well explains how they work.

- Scenario 1

Let us assume there is a blind person and two balls, one black and one white. You then would like to prove to the blind person that these balls are indeed of different colors without revealing the individual colors of each ball.

For this, you ask the blind person to hide both balls under the table and bring one ball back up for you to see.

After that, he should hide the ball back under the table and then either show the same ball or the other one. As a result, you can prove to the blind person that the colors are different by saying whether he changed the balls under the table or not.

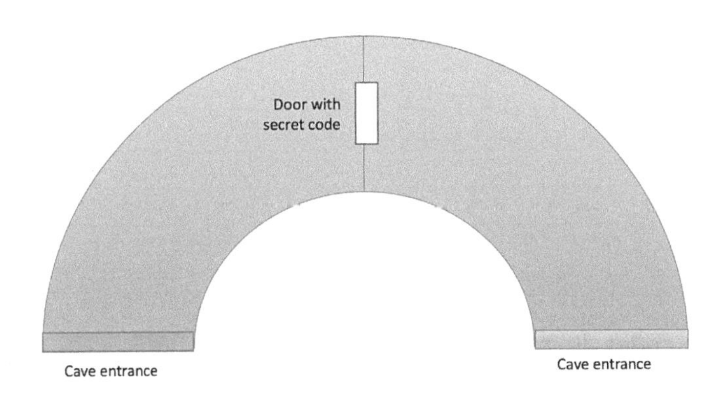

Figure 4.1: Cave example of ZKP

- Scenario 2

Pretend that there's a circular cave, with only one entrance or exit and at the back of this circular cave there's a door that can be unlocked using a secret code entered onto a keypad (Figure 4.1).

If I want to prove to you that I know the unlock code without revealing that unlock code to you, all I need to show is that I can walk into one end of the cave, open the door, and come out the other end.

If I have successfully demonstrated that, then you know without a doubt I have been able to unlock that door, but yet I have not revealed that unlock code to you.

4.2.2 Zero-knowledge proofs in Blockchain

Zero-knowledge protocols enable the transfer of assets across a distributed, peer-to-peer Blockchain network with complete privacy. In regular Blockchain transactions, when an asset is sent from one party to another, the details of that transaction are visible to every other party in the network.

By contrast, in a zero-knowledge transaction, the others only know that a valid transaction has taken place, but nothing about the sender, recipient, asset class and quantity. The identity and amount being spent can remain hidden. For example, a user may make a request to send another user some money.

The Blockchain naturally wants to make sure, before it commits this transaction, that the user who sends the money has enough money to send.

However, the Blockchain doesn't really need to know or care who is spending the money, or how much total money they have. Hyperledger Fabric and Ethereum have the implementation of Zero-Knowledge Proofs under development.

4.2.3 ZKP advantages

1. Zero-knowledge transfer as the name suggests

2. Computational efficiency- No Encryption
3. No degradation of the protocol

4.3 Sharding

Sharding: a solution to the scalability, latency and transaction throughput issues in Blockchain Sharding is a concept that is widely used in databases, to make them more efficient.

A shard is a horizontal portion of a database, with each shard stored in a separate server instance. This spreads the load and makes the database more efficient. In the case of the Blockchain, each node will have only a part of the data on the Blockchain, and not the entire information, when sharding is implemented.

Nodes that maintain a shard maintain information only on that shard in a shared manner, so within a shard, the decentralization is still maintained. However, each node does not load the information on the entire Blockchain, thus helping in scalability.

Blockchains that implement sharding use proof of stake (PoS) consensus algorithm.

4.3.1 What is Sharding in Blockchain?

In order to tackle the currently persisting issues with the validation mechanisms, a new kind of validation protocol has been created, i. e., Sharding. As part of Sharding only a small subset of nodes (called a *Shard*), out of the entire network nodes that will carry out validation of every single transaction. Sharding schema that is comprised of a cluster isolated Shards.

How a country gets divided into multiple states, in order to have a better governance system, the same way the Ethereum network will be logically divided into multiple Shards.

Transactions created by users or a particular Shard will be validated by miner's percent in that Shard alone.

4.3.2 Drawbacks of Sharding

If you think Sharding is the holy grail of all the scaling and performance issues, then you are mistaken. Sharding does come with its share of issues. The biggest flaw with Sharding is that inter Shard communication is not very easy.

What it basically means is that, as long as communication occurs with a Shard, the picture remains rosy and nice. But if a user (for e.g. Bob) who belongs to Shard-1 wants to transact with another user (say John) from Shard-2, the transaction would require some special protocols to complete the transaction.

The developer community would be the most affected lot, as they will have to program their codes to handle this.

4.5 What Is a Smart Contract?

A smart contract is a self-executing contract with the terms of the agreement between buyer and seller being directly written into lines of code. The code and the agreements contained therein exist across a distributed, decentralized Blockchain network. The code controls the execution, and transactions are trackable and irreversible. Smart contracts permit trusted transactions and agreements to be carried out among disparate, anonymous parties without the need for a central authority, legal system, or external enforcement mechanism. [13]

Smart contracts can be termed as the most utilized application of Blockchain technology in current times. The concept of smart contracts was introduced by Nick Szabo, a legal scholar, and cryptographer in the year 1994. He came to the conclusion that any decentralized ledger can be used as self-executable contracts which, later on, were termed as Smart Contracts. These digital contracts could be converted into codes and allowed to be run on a Blockchain.

Smart contracts are one of the most successful applications of the Blockchain technology. Using smart contracts in place of traditional ones can reduce the transaction costs significantly. Ethereum is the most popular Blockchain platform for creating smart contracts. It supports a feature called *Turing-completeness* that allows the creation of more customized smart contracts. Smart contracts can be applied in different industries and fields such as smart homes, e-commerce, real estate, and asset management, etc. [14]

Smart contracts are automatically executable lines of code that are stored on a Blockchain which contain predetermined rules (Figure 4.2). When these rules are met, these codes execute by themselves and provide the output. In the simplest form, smart contracts are programs that run according to the format that they've been set up by their creator. Smart contracts are most beneficial in business collaborations in which they are used to agree upon the decided terms set up by the consent of both the parties. This reduces the risk of fraud and as there is no third-party involved, the costs are reduced too. To summarize, smart contracts usually work on a mechanism that involves digital assets along with multiple parties where the involved participants can automatically govern their assets. These assets and be deposited and redistributed among the participants according to the rules of the contract. Smart contracts have the potential to track real-time performance and save costs.

Smart Contracts Properties:

- Self-verifiable
- Self-executable
- Tamper Proof

Example of Smart Contracts Code [15]

```
pragma solidity >=0.4.0 <0.7.0;

contract SimpleStorage {

        unit storedData;

        function set(unit x) public {

                storedData = x;

        }

        function get() public view returns (unit) {

                return storedData;

        }

}
```

Figure 4.2: Example of Smart Contract Code using Solidity (storage contract)

5

Decentralized Applications – DApps

Decentralized applications (DApps) are applications that run on a P2P (Peer-to-Peer) network of computers rather than a single computer. DApps, have existed since the advent of P2P networks. They are a type of software program designed to exist on the Internet in a way that is not controlled by any single entity. As opposed to simple smart contracts, in the classic sense of Bitcoin, which sends money from A to B, DApps have an *unlimited* number of participants on all sides of the market.

5.1 Difference between DApps & Smart Contracts

DApps are a 'Blockchain-enabled' website, where the Smart Contract is what allows it to connect to the Blockchain.

The easiest way to understand this is to understand how traditional websites operate. The traditional web application uses HTML, CSS, and JavaScript to render a page. It will also need to grab details from a database utilizing an API. When you go onto websites like Facebook, the page will call an API to grab your personal data and display them on the page (Figure 5.1).

Figure 5.1: Traditional Website Process

DApps are similar to a conventional web application. The front end uses the *exact same* technology to render the page. The one critical difference is that instead of an API connecting to a Database, you have a Smart Contract connecting to a Blockchain (Figure 5.2).

Figure 5.2: DApp enabled website

As opposed to traditional, centralized applications, where the back-end code is running on centralized servers, DApps have their backend code running on a decentralized P2P network.

Decentralized applications consist of the whole package, from backend to frontend.

But the smart contract is only one part of the DApp:

- Frontend (what you can see) and Backend (the logic on the background).
- A smart contract, on the other hand, consists only of the back-end, and often only a small part of the whole DApp.

That means if you want to create a decentralized application on a smart contract system, you have to combine several smart contracts and rely on 3rd party systems for the front-end.

DApps can have frontend code and user interfaces written in any language (just like an App) that can make calls to its backend. Furthermore, its frontend can be hosted on decentralized storage.

5.2 Blockchain DApps

For an application to be considered a DApp in the context of Blockchain, it must meet the following criteria:

1. Application must be completely open-source
   It must operate autonomously, and with no entity controlling the majority of its tokens. The application may adapt its protocol in response to proposed improvements and market feedback, but the consensus of its users must decide all changes.

2. Application's data and records of operation must be cryptographically stored
   Must be cryptographically stored in a public, decentralized Blockchain in order to avoid any central points of failure.

3. Application must use a cryptographic token
   (Bitcoin or a token native to its system) which is necessary for access to the application and any contribution of value from (miners) should be rewarded with the application's tokens.

4. Application must generate tokens
   According to a standard cryptographic algorithm acting as a proof of the value, nodes are contributing to the application (e.g.: Bitcoin uses the Proof of Work Algorithm).

6.3 Example: Ethereum DApps

Ethereum provides developers with a foundational layer: a Blockchain with a built-in Turing-complete programming language, allowing anyone to write smart contracts and decentralized applications where they can create their own arbitrary rules for ownership, transaction formats, and state transition functions.

In general, there are three types of applications on top of Ethereum.

1. Financial applications
   Providing users with more powerful ways of managing and entering into contracts using their money.

2. Semi-financial applications
   Where money is involved, but there is also a heavy non-monetary side to what is being done

3. Governance Applications

   Such as online voting & decentralized governance that are not financial at all.

Part 2

Blockchain Applications

6

Using Blockchain to Secure IoT

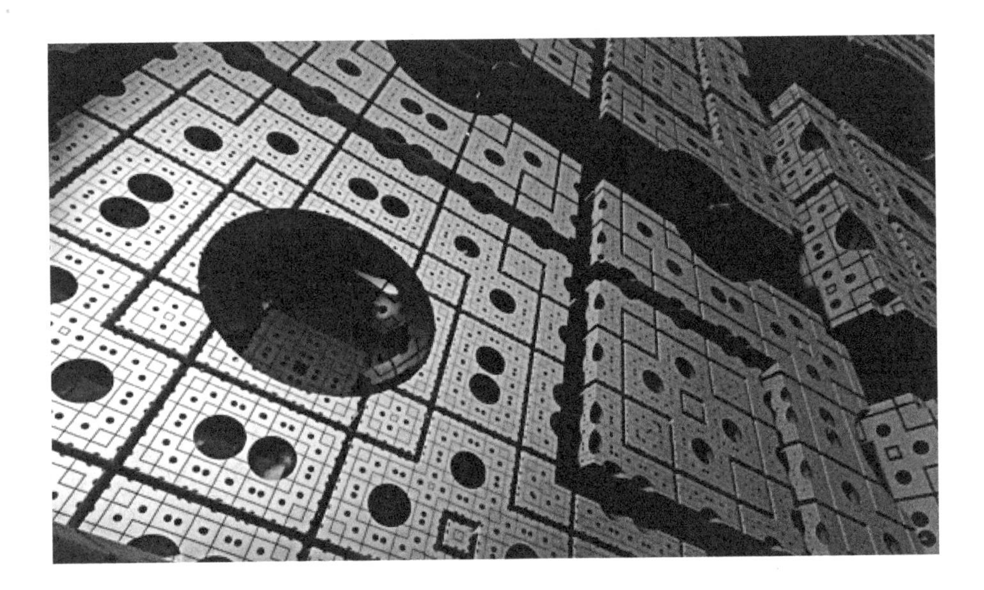

In an IoT world, information is the "fuel" that is used to change the physical state of environments through devices that are not general-purpose computers but, instead, devices and services that are designed for specific purposes. As such, the IoT is at a conspicuous inflection point for IT security. [16]

6.1 Challenges to Secure IoT Deployments

Regardless of the role, your business has within the Internet of Things ecosystem – device manufacturer, solution provider, cloud provider, systems integrator or service provider – you need to know how to get the greatest benefit from this new technology that offers such highly diverse and rapidly changing opportunities.

Handling the enormous volume of existing and projected data is daunting. Managing the inevitable complexities of connecting to a seemingly unlimited list of devices is complicated. And the goal of turning the deluge of data into valuable actions seems impossible due to the many challenges. The existing security technologies will play a role in mitigating IoT risks, but they are not enough. The goal is to get data securely at the right place, at the right time and in the right format, and it is easier said than done for many reasons.

6.2 Dealing with the Challenges and Threats

Gartner reported that more than 20% of businesses need to deploy security solutions for protecting their IoT devices and services. IoT devices and services will expand the surface area for cyber-attacks on businesses, by turning physical objects that used to be offline into online assets communicating with enterprise networks. Businesses will have to respond by broadening the scope of their security strategy to include these new online devices.

Businesses will have to tailor security to each IoT deployment according to the unique capabilities of the devices involved and the risks associated with the networks connected to those devices. BI Intelligence expects spending on solutions to secure IoT devices and systems to increase 5-fold over the next 4 years.

6.3 The Optimum Platform

Developing solutions for the Internet of Things requires unprecedented collaboration, coordination and connectivity for each piece in the system, and throughout the system as a whole. All devices must work together and be integrated with all other devices, and all devices must communicate and interact seamlessly with connected systems and infrastructures in a secure way. It is possible, but it can be expensive, time-consuming, and difficult unless the new line of thinking and a new approach to IoT security emerged away from the current centralized model.

The current IoT ecosystems rely on centralized, brokered communication models, otherwise known as the server/client paradigm. All devices are identified, authenticated and connected through cloud servers that sport huge processing and storage capacities. The connection between devices will have to exclusively go through the Internet, even if they happen to be a few feet apart.

While this model has connected generic computing devices for decades and will continue to support small-scale IoT networks as we see them today, it will not be able to respond to the growing needs of the huge IoT ecosystems of tomorrow.

Existing IoT solutions are expensive because of the high infrastructure and maintenance cost associated with centralized clouds, large server farms and networking equipment. The sheer amount of communications that will have to be handled when IoT devices grow to tens of billions will increase those costs substantially.

Even if the unprecedented economical and engineering challenges are overcome, cloud servers will remain a bottleneck and point of failure that can disrupt the entire network. This is especially important as more critical tasks.

Moreover, the diversity of ownership of devices and their supporting cloud infrastructure makes machine-to-machine (M2M) communications difficult. There is no single platform that connects all devices and no guarantee that cloud services offered by different manufacturers are interoperable and compatible. [17]

6.4 Decentralizing IoT Networks

A decentralized approach to IoT networking would solve many of the questions above. Adopting a standardized peer-to-peer communication model to process the hundreds of billions of transactions between devices will significantly reduce the costs associated with installing and maintaining large centralized data centers and will distribute computation and storage needs across the billions of devices that form IoT networks. This will prevent failure in any single node in a network from bringing the entire network to a halting collapse. [18]

However, establishing peer-to-peer communications will present its own set of challenges, chief among them the issue of security. And as we all know, IoT security is much more than just about protecting sensitive data. The proposed solution will have to maintain privacy and security in huge IoT networks and offer some form of validation and consensus for transactions to prevent spoofing and theft.

To perform the functions of traditional IoT solutions without a centralized control, any decentralized approach must support three fundamental functions:

- Peer-to-peer messaging
- Distributed file sharing
- Autonomous device coordination

6.5 The Blockchain Approach

Blockchain, the "distributed ledger" technology that underpins bitcoin, has emerged as an object of intense interest in the tech industry and beyond. Blockchain technology offers a way of recording transactions or any digital interaction in a way that is designed to be secure, transparent, highly resistant to outages, auditable and efficient; as such, it carries the possibility of disrupting industries and enabling new business models. The technology is young and changing very rapidly; widespread commercialization is still a few years off. Nonetheless, to avoid disruptive surprises or missed opportunities, strategists, planners and decision-makers across industries and business functions should pay heed now and begin to investigate applications of the technology. [19]

6.5.1 What is Blockchain?

Blockchain is a database that maintains a continuously growing set of data records. It is distributed in nature, meaning that there is no master computer holding the entire chain. Rather, the participating nodes have a copy of the chain. It is also ever-growing – data records are only added to the chain. [20]

A Blockchain consists of two types of elements:

- Transactions are the actions created by the participants in the system.
- Blocks record these transactions and ensure that they are in the correct sequence and have not been tampered with. Blocks also record a timestamp when the transactions were added.

6.5.2 What are Some Advantages of Blockchain?

There are three main advantages of Blockchain (Figure 6.1):

The big advantage of Blockchain is that it is *public*. Everyone participating can see the blocks and the transactions stored in them. This does not mean that everyone can see the actual content of your transaction; however, that is protected by your private key.

A Blockchain is *decentralized*, so there is no single authority that can approve the transactions or set specific rules to have transactions accepted. That means there is a huge amount of trust involved since all the participants in the network have to reach a consensus to accept transactions.

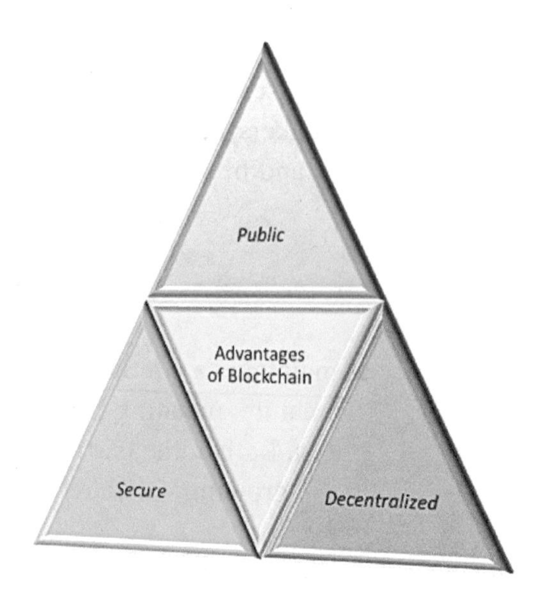

Figure 6.1: Advantages of Blockchain

Most importantly, it is *secure*. The database can only be extended and previous records cannot be changed (at least, there is a very high cost if someone wants to alter previous records).

6.5.3 How Does It Work?

When someone wants to add a transaction to the chain, all the participants in the network will validate it. They do this by applying an algorithm to the transaction to verify its validity. What exactly is understood by "valid" is defined by the Blockchain system and can differ between the systems. Then, it is up to the majority of the participants to agree that the transaction is valid.

A set of approved transactions is then bundled in a block, which is then sent to all the nodes in the network. They, in turn, validate the new block. Each successive block contains a hash, which is a unique fingerprint, of the previous block. [13]

6.6 The Blockchain and IoT

Blockchain technology is the missing link to settle scalability, privacy and reliability concerns in the Internet of Things. Blockchain technologies could perhaps be the silver bullet needed by the IoT industry. Blockchain technology can be used in tracking billions of connected devices, enable the processing of transactions and coordination between devices and allow for significant savings to IoT industry manufacturers. This decentralized approach would eliminate single points of failure, creating a more resilient ecosystem for devices to run on. The cryptographic algorithms used by Blockchains would make consumer data more private.

The ledger is tamper-proof and cannot be manipulated by malicious actors because it does not exist in any single location, and man-in-the-middle attacks cannot be staged because there is no single thread of communication that can be intercepted. Blockchain makes trustless, peer-to-peer messaging possible and has already proven its worth in the world of financial services through cryptocurrencies such as Bitcoin, providing guaranteed peer-to-peer payment services without the need for third-party brokers.

The decentralized, autonomous and trustless capabilities of the Blockchain make it an ideal component to become a fundamental element of IoT solutions. It is not a surprise that enterprise IoT technologies have quickly become one of the early adopters of Blockchain technologies.

In an IoT network, the Blockchain can keep an immutable record of the history of smart devices. This feature enables the autonomous functioning of smart devices without the need for centralized authority. As a result, the Blockchain opens the door to a series of IoT scenarios that were remarkably difficult, or even impossible to implement without it.

By leveraging the Blockchain, IoT solutions can enable secure, trustless messaging between devices in an IoT network. In this model, the Blockchain will treat message exchanges between devices similar to financial transactions in a bitcoin network. To enable message exchanges, devices will leverage smart contracts, which then model the agreement between the two parties.

In this scenario, we can sensor from afar, communicating directly with the irrigation system in order to control the flow of water-based on conditions detected on the crops. Similarly, smart devices in an oil platform can exchange data to adjust functioning based on weather conditions.

Using the Blockchain will enable true autonomous smart devices that can exchange data, or even execute financial transactions, without the need of a centralized broker. This type of autonomy is possible because the nodes in the Blockchain network will verify the validity of the transaction without relying on a centralized authority.

In this scenario, we can envision smart devices in a manufacturing plant that can place orders for repairing some of its parts without the need for human or centralized intervention. Similarly, smart vehicles in a truck fleet will be able to provide a complete report of the most important parts needing replacement after arriving at a workshop.

One of the most exciting capabilities of the Blockchain is the ability to maintain a duly decentralized, trusted ledger of all transactions occurring in a network. This capability is essential to enable the many compliances and regulatory requirements of industrial IoT applications without the need to rely on a centralized model. [15] [21]

7

IoT and Blockchain: Challenges and Risks

The Internet of Things (IoT) is an ecosystem of ever-increasing complexity; it is the next wave of innovation that will humanize every object in our life, and it is the next level of automation for every object we use. IoT is bringing more and more things into the digital fold every day, which will likely make IoT a multi-trillion-dollar industry in the near future. To understand the scale of interest in the Internet of Things (IoT), just check how many conferences, articles, and studies have been conducted about IoT recently. This interest has hit fever pitch point in 2016, as many companies see big opportunity and believe that IoT holds the promise to expand and improve business processes and accelerate growth. However, the rapid evolution of the IoT market has caused an explosion in the number and variety of IoT solutions, which created real challenges as the industry evolves, mainly the urgent need for a secure IoT model to perform common tasks such as sensing, processing, storage and communicating. Developing that model will never be an easy task by any stretch of the imagination, and there are many hurdles and challenges facing a real secure IoT model. [13]

The biggest challenge facing IoT security is coming from the very architecture of the current IoT ecosystem; it is all based on a centralized model known as the server/client model. All devices are identified, authenticated and connected through cloud servers that support huge processing and storage capacities. The connection between devices will have to go through the cloud, even if they happen to be a few feet apart. While this model has connected computing devices for decades and will continue to support today's IoT networks, it will not be able to respond to the growing needs of the huge IoT ecosystems of tomorrow.

7.1 The Blockchain Model

The Blockchain is a database that maintains a continuously growing set of data records. It is distributed in nature, meaning that there is no master computer holding the entire chain. Rather, the participating nodes have a copy of the chain. It is also ever-growing – data records are only added to the chain.

When someone wants to add a transaction to the chain, all the participants in the network will validate it. They do this by applying an algorithm to the transaction to verify its validity. What exactly is understood by "valid" is defined by the Blockchain system and can differ between the systems. Then, it is up to the majority of the participants to agree that the transaction is valid.

A set of approved transactions is then bundled in a block, which is sent to all the nodes in the network. They, in turn, validate the new block. Each successive block contains a hash, which is a unique fingerprint, of the previous block. [22]

7.2 Principles of Blockchain Technology

Here are five basic principles underlying the technology. [23]

1. Distributed Database
   Each party on a Blockchain has access to the entire database and its complete history. No single party controls the data or the information. Every party can verify the records of its transaction partners directly, without an intermediary.

2. Peer-to-Peer Transmission
   Communication occurs directly between peers instead of through a central node. Each node stores and transfers pieces of information to all other nodes.

3. Transparency
   Every transaction and its associated value are visible to anyone with access to the system. Each node, or user, on a Blockchain, has a unique 30-plus-character alphanumeric address that identifies it. Users can choose to remain anonymous or provide proof of their identity to others. Transactions occur between Blockchain addresses.

4. Irreversibility of Records
   Once a transaction is entered in the database and the accounts are updated, the records cannot be altered, because they are linked to every transaction record that came before them (hence the term "chain"). Various computational algorithms and approaches are deployed to ensure that the recording on the database is permanent, chronologically ordered and available to all others on the network.

5. Computational Logic
   The digital nature of the ledger means that Blockchain transactions can be tied to computational logic and in essence programmed. Therefore, users can set up algorithms and rules that automatically trigger transactions between nodes.

7.3 Challenges of Blockchain in IoT

In spite of all its benefits, the Blockchain model is not without flaws and shortcomings, (Figure 7.1) which are presented below: [19]

Scalability issues related to the size of Blockchain ledger that might lead to centralization as it is grown over time and require some kind of

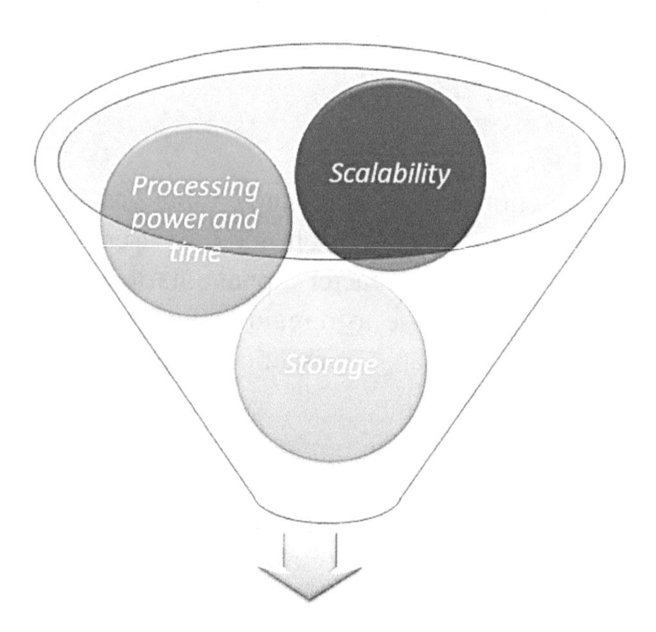

Figure 7.1: Challenges of Blockchain In IoT

record management, which is casting a shadow over the future of the Blockchain technology.

Processing power and time required to perform encryption algorithms for all the objects involved in Blockchain-based IoT ecosystem given the fact that IoT ecosystems are very diverse and composed of devices that have very different computing capabilities, and not all of them will be capable of running the same encryption algorithms at the desired speed.

Storage will be a hurdle: Blockchain eliminates the need for a central server to store transactions and device IDs, but the ledger has to be stored on the nodes themselves, and the ledger will increase in size as time passes. That is beyond the capabilities of a wide range of smart devices such as sensors, which have very low storage capacity.

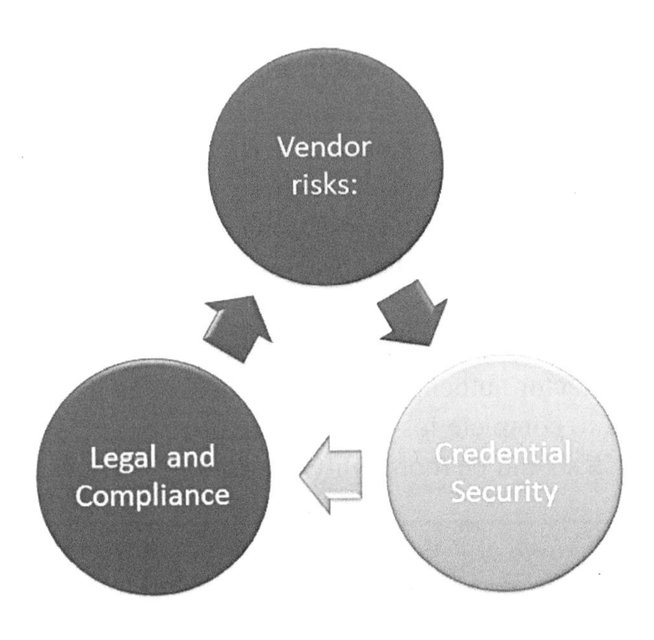

Figure 7.2: Risks of Using Blockchain in IoT

7.4 Risks of Using Blockchain in IoT

It goes without saying that any new technology comes with new risks. An organization's risk management team should analyze, assess and design mitigation plans for risks expected to emerge from the implementation of Blockchain-based frameworks (Figure 7.2).

Vendor Risks: Practically speaking, most present organizations, looking to deploy Blockchain-based applications, lack the required technical skills and expertise to design and deploy a Blockchain-based system and implement smart contracts completely in-house, i.e. without reaching out for vendors of Blockchain applications. The value of these applications is only as strong as the credibility of the vendors providing them. Given the fact that the Blockchain-as-a-Service (BaaS) market is still a developing market, a business should meticulously select a vendor that can perfectly sculpture applications that appropriately address the risks associated with the Blockchain.

Credential Security: Even though the Blockchain is known for its high-security levels, a Blockchain-based system is only as secure as the system's access point. When considering a public Blockchain-based system, any individual who has access to the private key of a given user, which enables him/her to "sign" transactions on the public ledger, will effectively become that user, because most current systems do not provide multi-factor authentication. Also, loss of an account's private keys can lead to complete loss of funds, or data, controlled by this account; this risk should be thoroughly assessed.

Legal and Compliance: It is a new territory in all aspects without any legal or compliance precedents to follow, which poses a serious problem for IoT manufacturers and service providers. This challenge alone will scare off many businesses from using Blockchain technology.

7.5 The Optimum Secure IoT Model

In order for us to achieve that optimal secure model of IoT, security needs to be built in as the foundation of IoT ecosystem, with rigorous validity checks, authentication, data verification, and all the data need to be encrypted at all levels, without a solid bottom-top structure, and we will create more threats with every device added to the IoT. What we need is a secure and safe IoT with privacy protected. That is a tough trade-off, but possible with Blockchain technology if we can overcome its drawbacks. [24]

8

IoT, AI, and Blockchain: Catalysts for Digital Transformation

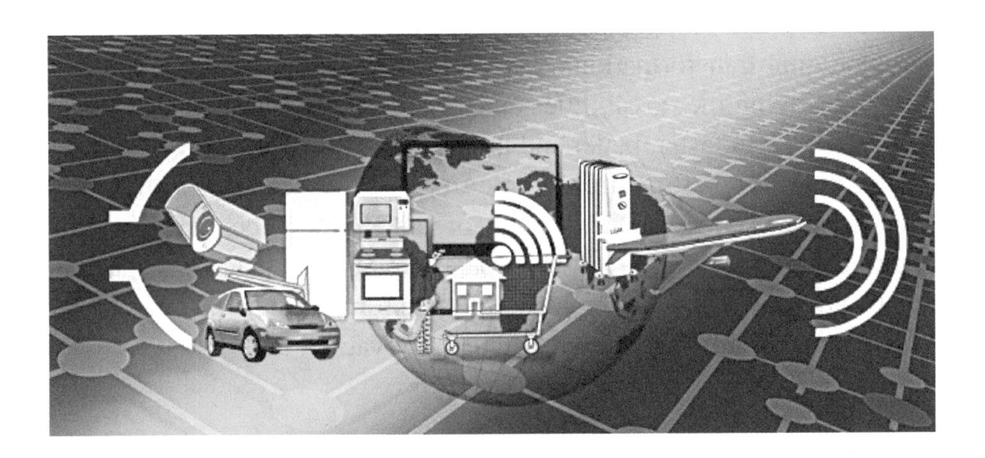

The digital revolution has brought with it a new way of thinking about manufacturing and operations. Emerging challenges associated with logistics and energy costs are influencing global production and associated distribution decisions. Significant advances in technology, including Big Data analytics, AI, Internet of Things, robotics and additive manufacturing, are shifting the capabilities and value proposition of global manufacturing. In response, manufacturing and operations require a digital renovation: the value chain must be redesigned and retooled and the workforce retrained. Total delivered cost must be analyzed to determine the best places to locate sources of supply, manufacturing and assembly operations around the world. In other words, we need a digital transformation.

8.1 Digital Transformation

Digital transformation (DX) is the profound transformation of business and organizational activities, processes, competencies, and models to fully leverage the changes and opportunities of a mix of digital technologies and their accelerating impact across society in a strategic and prioritized way, with present and future shifts in mind (Figure 8.1).

A digital transformation strategy aims to create the capabilities of fully leveraging the possibilities and opportunities of new technologies and their impact faster, better and in more innovative ways in the future.

A digital transformation journey needs a staged approach with a clear road-map, involving a variety of stakeholders, beyond silos and internal/external limitations. This road-map takes into account that end goals will continue to move as digital transformation *de facto* is an ongoing journey, as is change and digital innovation. [25]

8.2 Internet of Things (IoT)

IoT is defined as a system of interrelated *physical objects, sensors, actuators, virtual objects, people, services, platforms and networks* that have separate identifiers and an ability to transfer data independently. Practical examples of IoT applications today include precision agriculture, remote patient monitoring, and driverless cars. Simply put, IoT is the network of "things" that collects and exchanges information from the environment.

IoT and digital transformation are closely related to the following reasons: [26] [27]

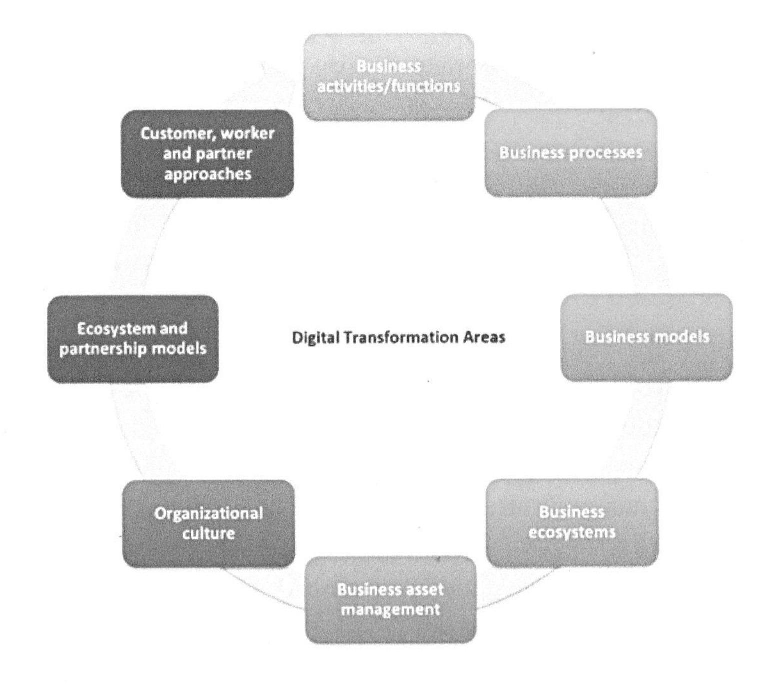

Figure 8.1: Digital Transformation Areas

1. More than 50% of companies think IoT is strategic, and one in four believes it is transformational.
2. Both increase company longevity. The average life span of a company has decreased from 67 years in the 1920s to 15 years today.
3. One in three industry leaders will be digitally disrupted.
4. Both enable businesses to connect with customers and partners in open digital ecosystems, share digital insights, collaborate on solutions and share in the value created.
5. Competitors are doing it. According to IDC, 70% of global discrete manufacturers will offer connected products by 2020.

6. It is where the money is. Digital product and service sales are growing and will represent more than US $1 of every US $3 spent by 2021.
7. Enterprises are overwhelmed by data and digital assets. They already struggle to manage the data and digital assets they have, and IoT will expand them exponentially. They need help finding insights into the vast stream of data and manage digital assets.
8. Both drive consumption. Digital services easily prove their own worth. Bundle products with digital services and content make it easy for customers to consume them.
9. Both make companies understand customers better. Use integrated channels, Big Data, predictive analytics, and machine learning to uncover, predict and meet customer needs, increasing loyalty and revenues, IoT and AI are at the heart of this.
10. Using both is future-proof for the business. Make the right strategic bets for the company, product and service portfolio and future investments using IoT data analytics, visualization, and AI.

8.3 Digital Transformation, Blockchain, and AI

Digital transformation is a complicated challenge, but the integration of Blockchain and AI makes it much easier. Considering the number of partners (internal, external or both) involved in any given business process, a system in which a multitude of electronic parties can securely communicate, collaborate and transact without human intervention is highly agile and efficient.

Enterprises that embrace this transformation will be able to provide a better user experience, a more consistent workflow, more streamlined operations, and value-added services, as well as gain competitive advantage and differentiation.

Blockchain can holistically manage steps and relationships where participants will share the same data source, such as financial relationships and transactions connected to each step, security and accountability factored in, as well as compliance with government regulations along with internal rules and processes. The result is consistency, reductions in costs and time delays, improved quality and reduced risks. [28]

AI can help companies learn in ways that accelerate innovation and assist companies in getting closer to customers and improve employee's productivity and engagement. Digital transformation efforts can be improved with that information.

8.4 Conclusion

The building blocks of digital transformation are mindset, people, process and tools. IoT covers all the blocks since IoT does not just connect devices, it connects people too. Blockchain will ensure end-to-end security, and by using AI, you will move IoT beyond connections to intelligence. One important step is to team up with the best partners and invest in education, training and certifying your teams. This magical mix of IoT, AI, and Blockchain will help make transformation digital and easy. [29]

9

Myths about Blockchain Technology

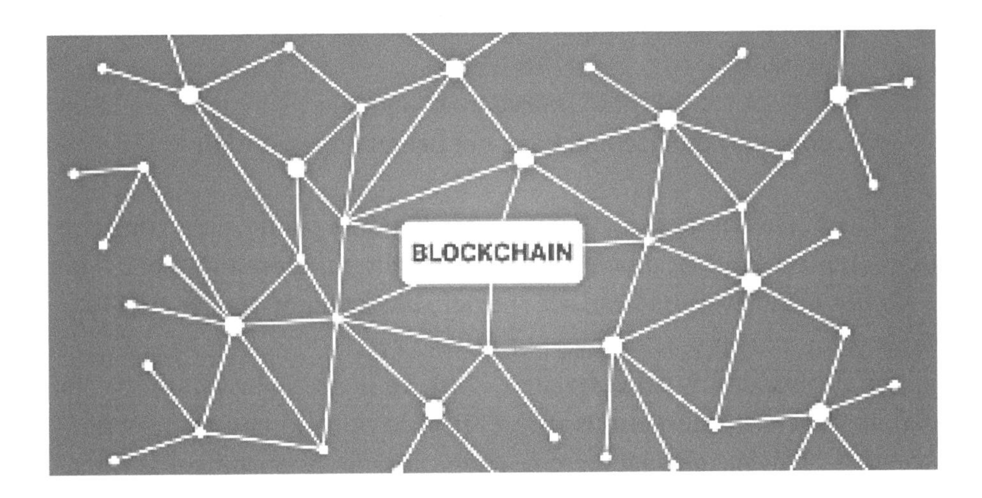

Blockchain, the "distributed ledger" technology, has emerged as an object of intense interest in the tech industry and beyond. Blockchain technology offers a way of recording transactions or any digital interaction in a way that is designed to be secure, transparent, highly resistant to outages, auditable and efficient; as such, it carries the possibility of disrupting industries and enabling new business models. The technology is young and changing very rapidly; widespread commercialization is still a few years off. Nonetheless, to avoid disruptive surprises or missed opportunities, strategists, planners and decision-makers across industries and business functions should pay heed now and begin to investigate applications of the technology.

Blockchain is a database that maintains a continuously growing set of data records. It is distributed in nature, meaning that there is no master computer holding the entire chain. Rather, the participating nodes have a copy of the chain. It is also ever-growing – data records are only added to the chain.

A Blockchain consists of two types of elements: [30]

- Transactions are the actions created by the participants in the system.
- Blocks record these transactions and make sure they are in the correct sequence and have not been tampered with.

The big advantage of Blockchain is that it is public. Everyone participating can see the blocks and the transactions stored in them. This does not mean everyone can see the actual content of your transaction, however; that is protected by your private key.

A Blockchain is decentralized, so there is no single authority that can approve the transactions or set specific rules to have transactions accepted. This means that there is a huge amount of trust involved since all the participants in the network have to reach a consensus to accept transactions.

Most importantly, it is secure. The database can only be extended and previous records cannot be changed (at least, there is a very high cost if someone wants to alter previous records).

When someone wants to add a transaction to the chain, all the participants in the network will validate it. They do this by applying an algorithm to the transaction to verify its validity. What exactly is understood by "valid" is defined by the Blockchain system and can differ between the systems. Then, it is up to the majority of the participants to agree that the transaction is valid.

A set of approved transactions is then bundled in a block, which is then sent to all the nodes in the network. They, in turn, validate the new block. Each successive block contains a hash, which is a unique fingerprint, of the previous block.

Blockchain ensures that data has not been tampered with, offering a layer of time-stamping that removes multiple levels of human checking and makes transactions immutable. However, it is not yet the cure-all that some believe it to be. [97]

Blockchain technology certainly has many positive aspects, but there is also much misunderstanding and confusion regarding its nature.

9.1 Myth 1: The Blockchain Is a Magical Database in the Cloud [31]

The Blockchain is conceptually a flat-file – a linear list of simple transaction records. "This list is appended only so entries are never deleted, but instead, the file grows indefinitely and must be replicated in every node in the peer-to-peer network."

Blockchain does not allow you to store any type of physical information like a Word document or a pdf file. It can only provide a "proof of existence", and the distributed ledger can only contain a code that certifies the existence of a certain document but not the document itself. The file, however, can be stored in "data lakes", the access to which is controlled by the owner of the information.

9.2 Myth 2: Blockchain Is Going to Change the World [32]

We can use Blockchain for complex and technical transactions such as verifying the authenticity of a diamond or the identity of a person. There is also talk of a Blockchain application for the bill of lading in trade finance, which would be revolutionary in terms of cost reduction and transaction speed.

While Blockchain can support these cases and mitigate the risk of a fraudster tampering with the ledger, it does not eradicate the threat of fraud online and it still raises questions over confidentiality. Additionally, the use of Blockchain technology will still be inefficient for many of these cases when compared to maintaining a traditional ledger.

9.3 Myth 3: Blockchain Is Free [28]

Despite the commonly held belief, Blockchain is neither cheap nor efficient to run. However, it involves multiple computers solving mathematical algorithms to agree on a final immutable result, which becomes the so-called single version of truth (SVT). Each "block" in the Blockchain typically uses a large amount of computing power to solve. And someone needs to pay for all this computer power that supports the Blockchain service.

9.4 Myth 4: There Is Only One Blockchain [28]

There are many different technologies that go by the name Blockchain. They come in public and private versions, open and closed source, general-purpose and tailored to specific solutions.

The common denominator is that they are sheared up by crypto, are distributed and have some form of consensus mechanism. Bitcoin's Blockchain, Ethereum, Hyperledger, Corda, and IBM and Microsoft's Blockchain-as-a-Service can all be classified as distributed ledger technologies.

9.5 Myth 5: The Blockchain Can Be Used for Anything and Everything [33]

Although the code is powerful, it is not magical. Bitcoin and Blockchain developers can be evangelical, and it is easy to understand why. For many, the Blockchain is an authority tied to mathematics, not the government or lawyers. In the minds of some developers, the Blockchain and smart contracts will one day replace money, lawyers and other arbitration bodies. Yet the code is limited to the number of cryptocurrency transactions in the chain itself, and cryptocurrency is still far from mainstream.

9.6 Myth 6: The Blockchain Can Be the Backbone of a Global Economy [29]

No national or corporate entity owns or controls the Blockchain. For this reason, evangelists hope that private Blockchains can provide foundational support for dozens of encrypted and trusted cryptocurrencies. Superficially, the Bitcoin Blockchain appears massive. Yet a Gartner report has recently claimed that the size of the Blockchain is similar in scale to the NASDAQ network. If cryptocurrency takes off, and records are generated larger, this may change. For now, the Blockchain network is roughly analogous to contemporary financial networks.

9.7 Myth 7: The Blockchain Ledger Is Locked and Irrevocable

Analogous large-scale transaction databases like bank records are, by their nature, private and tied to specific financial institutions. The power of Blockchain, of course, is that the code is public, transactions are verifiable and the network is cryptographically secure. Fraudulent transactions – double spends, in industry parlance – are rejected by the network, preventing fraud. Because mining the chain provides financial incentives in the form of Bitcoin, it is largely believed that rewriting historic transactions is not in the financial interest of participants. For now, however, as computational resources improve with time, so too does the potential for deception. The impact of future processing power on the integrity of the contemporary Blockchain remains unclear.

9.8 Myth 8: Blockchain Records Can Never Be Hacked or Altered [34]

One of the main selling points about Blockchains is their inherent permanence and transparency. When people hear that, they often think that Blockchains are invulnerable to outside attacks. No system or database will ever be completely secure, but the larger and more distributed the network, the more secure it is believed to be. What Blockchains can provide to applications that are developed on top of them is a way of catching unauthorized changes to records.

9.9 Myth 9: Blockchain Can Only Be Used in the Financial Sector [35]

Blockchain started to create waves in the financial sector because of its first application, the bitcoin cryptocurrency, which directly impacted this field. Although Blockchain has numerous areas of application, finance is undeniably one of them. The important challenges that this technology brings to the financial world pushed international banks such as Goldman Sachs or Barclays to heavily invest in it. Outside the financial sector, Blockchain can and will be used in real estate, healthcare or even at a personal scale to create a digital identity. Individuals could potentially store proof of the existence of medical data on the Blockchain and provide access to pharmaceutical companies in exchange for money.

9.10 Myth 10: Blockchain is Bitcoin [31]

Since Bitcoin is more famous than the underlying technology, Blockchain, many people get confused between the two.

Blockchain is a technology that allows peer-to-peer transactions to be recorded on a distributed ledger across the network. These transactions are stored in blocks and each block is linked to the previous one, therefore creating a chain. Thus, each block contains a complete and time-stamped record of all the transactions that occurred in the network. On the Blockchain, everything is transparent and permanent. No one can change or remove a transaction from the ledger.

Bitcoin is a cryptocurrency that makes electronic payments possible directly between two people without going through a third party like a bank. Bitcoins are created and stored in a virtual wallet. Since there are no intermediaries between the two parties, no one can control the cryptocurrency. Hence, the number of bitcoins that will ever be released is limited and defined by a mathematical algorithm.

9.11 Myth 11: Blockchain Is Designed for Business Interactions Only [31]

Experts in Blockchain are convinced that this technology will change the world and the global economy just like dot-coms did in the early 1990s. Hence, it is not only open to big corporations but is also accessible to everyone everywhere. If all it takes is an Internet connection to use the Blockchain, one can easily imagine how many people worldwide will be able to interact with each other.

9.12 Myth 12: Smart Contracts Have the Same Legal Value as Regular Contracts

For now, smart contracts are just pieces of code that execute actions automatically when certain conditions are met. Therefore, they are not considered as regular contracts from a legal perspective. However, they can be used as proof of whether or not a certain task has been accomplished. Despite their uncertain legal value, smart contracts are very powerful tools, especially when combined with the Internet of Things (IoT).

10

Cybersecurity & Blockchain

With the fact that cybercrime and cybersecurity attacks hardly seem to be out of the news these days and the threat is growing globally, nobody would appear immune to malicious and offensive acts targeting computer networks, infrastructures, and personal computer devices. Firms must clearly invest to stay resilient. Gauging the exact size of cybercrime and putting a precise US dollar value on it is nonetheless tricky. But one thing we can be sure about is that the number is big and probably larger than the statistics reveal.

The global figure for cyber breaches had been put at around US $200 billion annually [36]. Malicious cyber activity cost the U.S. economy between $57 billion and $109 billion in 2016, the White House Council of Economic Advisers estimated in a report released in February of 2018. [37]

New Blockchain platforms are stepping up to address security concerns in the face of recent breaches. Since these platforms are not controlled by a singular entity, they can help ease the concerns created by a spree of recent breach disclosures. Services built on top of Blockchain have the potential to inspire renewed trust due to the transparency built into the technology.

Developments in Blockchain have expanded beyond recordkeeping and cryptocurrencies. The integration of *smart contract* development in Blockchain platforms has ushered in a wider set of applications, including cybersecurity.

By using Blockchain, transaction details are kept both transparent and secure. Blockchain's decentralized and distributed network also helps businesses to avoid a single point of failure, making it difficult for malicious parties to steal or tamper with business data.

Transactions in the Blockchain can be audited and traced. In addition, public Blockchains rely on distributed network to run, thus eliminating a single point of control. For attackers, it is much more difficult to attack a large number of peers distributed globally as opposed to a centralized data center.

10.1 Implementing Blockchain in Cybersecurity

Since a Blockchain system is protected with the help of ledgers and cryptographic keys, attacking and manipulating it becomes extremely difficult. Blockchain decentralizes the systems by distributing ledger data on several systems rather than storing them on one single network. This allows the technology to focus on gathering data rather than worrying about any data being stolen. Thus, decentralization has led to an improved efficiency in Blockchain-operated systems.

For a Blockchain system to be penetrated, the attacker must intrude into every system on the network to manipulate the data that is stored on the network. The number of systems stored on every network can be in millions. Since domain editing rights are only given to those who require them, the hacker will not get the right to edit and manipulate the data even after hacking a million systems. Since such manipulation of data on the network has never taken place on the Blockchain, it is not an easy task for any attacker.

While we store our data on a Blockchain system, the threat of a possible hack gets eliminated. Every time our data is stored or inserted into Blockchain ledgers; a new block is created. This block further stores a key that is cryptographically created. This key becomes the unlocking key for the next record that is to be stored onto the ledger. In this manner, the data is extremely secure.

Furthermore, the hashing feature of Blockchain technology [38] is one of its underlying qualities that makes it such a prominent technology. Using cryptography and the hashing algorithm, Blockchain technology converts the data stored in our ledgers. This hash encrypts the data and stores it in such a language that the data can only be decrypted using keys stored in the systems. Other than cybersecurity, Blockchain has many applications in several fields that help in maintaining and securing data. The fields where this technology is already showing its ability are finance, supply chain management and Blockchain-enabled smart contracts. [39]

10.2 Advantages of Using Blockchain in Cybersecurity

The main advantages of Blockchain technology in cybersecurity (Figure 10.1) are the following: [40] [41] [42] [43]

10.2.1 Decentralization

Thanks to the peer-to-peer network, there is no need for third-party verification, as any user can see network transactions.

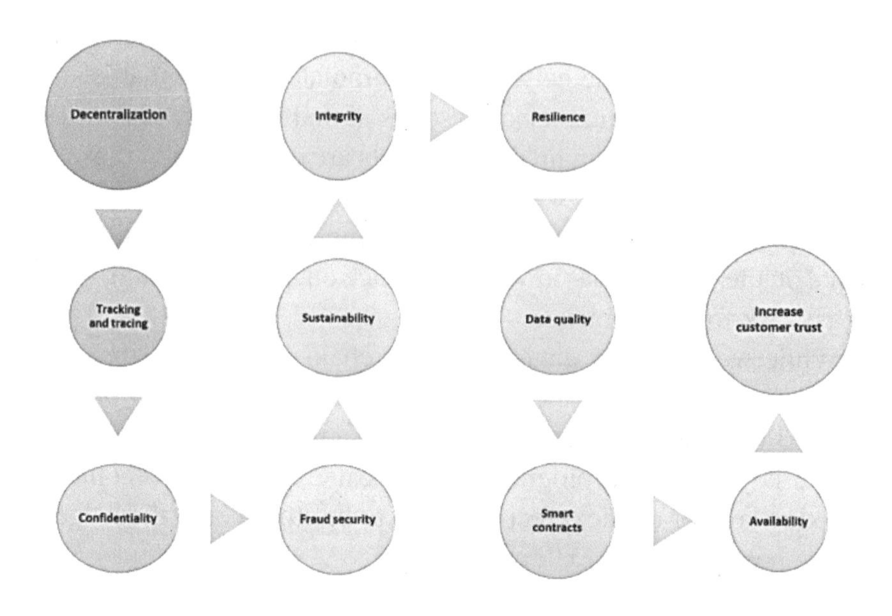

Figure 10.1: Advantages of Using Blockchain in Cybersecurity

10.2.2 Tracking and Tracing

All transactions in Blockchains are digitally signed and time-stamped, so network users can easily trace the history of transactions and track accounts at any historical moment. This feature also allows a company to have valid information about assets or product distribution.

11.2.3 Confidentiality

The confidentiality of network members is high due to the public-key cryptography that authenticates users and encrypts their transactions.

10.2.4 Fraud Security

In the event of a hack, it is easy to define malicious behavior due to the peer-to-peer connections and distributed consensus. As of today, Blockchains are considered technically "unhackable", as attackers can impact a network only by getting control of 51% of the network nodes.

11.2.5 Sustainability

Blockchain technology has no single point of failure, which means that even in the case of DDoS attacks, the system will operate as normal, thanks to multiple copies of the ledger.

10.2.6 Integrity

The distributed ledger ensures the protection of data against modification or destruction. Besides, the technology ensures the authenticity and irreversibility of completed transactions. Encrypted blocks contain immutable data that is resistant to hacking.

10.2.7 Resilience

The peer-to-peer nature of the technology ensures that the network will operate round-the-clock even if some nodes are offline or under attack. In the event of an attack, a company can make certain nodes redundant and operate as usual.

10.2.8 Data Quality

Blockchain technology cannot improve the quality of your data, but it can guarantee the accuracy and quality of data after it is encrypted in the Blockchain.

10.2.9 Smart Contracts

These are software programs that are based on the ledger. These programs ensure the execution of contract terms and verify parties. Blockchain technology can significantly increase the security standards for smart contracts, as it minimizes the risks of cyber-attacks and bugs.

10.2.10 Availability

There is no need to store your sensitive data in one place, as Blockchain technology allows you to have multiple copies of your data that are always available to network users.

10.2.11 Increase Customer Trust

Your clients will trust you more if you can ensure a high level of data security. Moreover, Blockchain technology allows you to provide your clients with information about your products and services instantly.

10.3 Disadvantages of Using Blockchain in Cybersecurity (Figure 10.2): [36] [37] [39]

10.3.1 Irreversibility

There is a risk that encrypted data may be unrecoverable in case a user loses or forgets the private key necessary to decrypt it.

10.3.2 Storage Limits

Each block can contain no more than 1 Mb of data, and a Blockchain can handle only seven transactions per second on average.

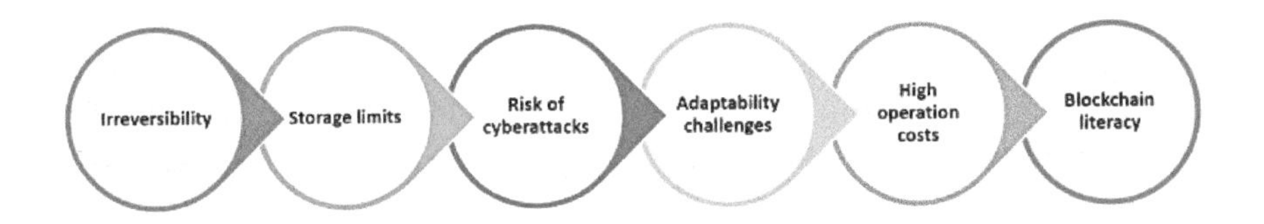

Figure 10.2: Disadvantages of Using Blockchain in Cybersecurity

10.3.3 Risk of Cyberattacks

Although the technology greatly reduces the risk of malicious intervention, it is still not a panacea to all cyber-threats. If attackers manage to exploit the majority of your network, you may lose your entire database.

10.3.4 Adaptability Challenges

Although Blockchain technology can be applied to almost any business, companies may face difficulties integrating it. Blockchain applications can also require complete replacement of existing systems, so companies should consider this before implementing the Blockchain technology.

10.3.5 High Operation Costs

Running Blockchain technology requires substantial computing power, which may lead to high marginal costs in comparison to existing systems.

10.3.6 Blockchain Literacy

There are still not enough developers with experience in Blockchain technology and with deep knowledge of cryptography.

10.4 Conclusion

Blockchain's decentralized approach to cybersecurity can be seen as a fresh take on the issues that the industry faces today. The market could only use more solutions to combat the threats of cyberattacks. And the use of Blockchain may yet address the vulnerabilities and limitations of current security approaches and solutions.

Throwing constant pots of money at the problem and knee-jerk reactions is not the answer. Firms need to sort out their governance, awareness and organizational culture and critically look at the business purpose and processes before they invest in systems to combat cybercrime.

The roster of these new services provided by Blockchain may be limited for now and of course, they face incumbent players in the cybersecurity space. But this only offers further opportunities for other ventures to cover other key areas of cybersecurity. Blockchain also transcends borders and nationalities, which should inspire trust in users. And, with the growth of these new solutions, the industry may yet restore some of the public's trust they may have lost in the midst of all these issues.

Overall, Blockchain technology is a breakthrough in cybersecurity, as it can ensure the highest level of data confidentiality, availability and security. However, the complexity of the technology may cause difficulties with development and real-world use.

Implementation of Blockchain applications requires comprehensive, enterprise- and risk-based approaches that capitalize on cybersecurity risk frameworks, best practices, and cybersecurity assurance services to mitigate risks. In addition, cyber intelligence capabilities, such as cognitive security, threat modeling, and artificial intelligence, can help to proactively predict cyber threats to create countermeasures. That is why AI is considered as the first line of defense while Blockchain is the second line. [44]

11

Blockchain and AI: A Perfect Match?

Blockchain and Artificial Intelligence are two of the hottest technology trends right now. Even though the two technologies have highly different developing parties and applications, researchers have been discussing and exploring their combination. [45]

PwC predicts that by 2030 AI will add up to $15.7 trillion to the world economy, and as a result, global GDP will rise by 14%. According to Gartner's prediction, the business value added by blockchain technology will increase to $3.1 trillion by the same year.

By definition, a blockchain is a distributed, decentralized, immutable ledger used to store encrypted data. On the other hand, AI is the engine or the "brain" that will enable analytics and decision making from the data collected. [46]

It goes without saying that each technology has its own individual degree of complexity, but both AI and blockchain are in situations where they can benefit from each other, and help one another. [47]

With both these technologies able to effect and enact upon data in different ways, their coming together makes sense, and it can take the exploitation of data to new levels. At the same time, the integration of machine learning and AI into the blockchain, and vice versa, can enhance blockchain's underlying architecture and boost AI's potential.

Additionally, blockchain can also make AI more coherent and understandable, and we can trace and determine why decisions are made in machine learning. Blockchain and its ledger can record all data and variables that go through a decision made under machine learning.

Moreover, AI can boost blockchain efficiency far better than humans, or even standard computing can. A look at the way in which blockchains are currently run on standard computers proves this with a lot of processing power needed to perform even basic tasks (Figure 11.1). [43]

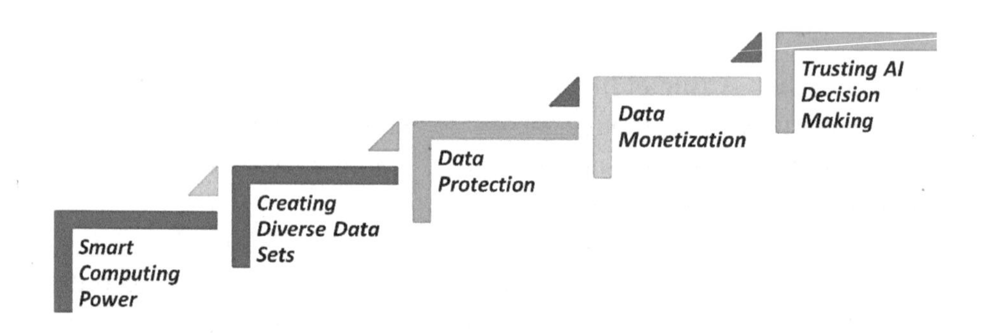

Figure 11.1: Applications of AI and Blockchain

Smart Computing Power

If you were to operate a blockchain, with all its encrypted data, on a computer you'd need large amounts of processing power. The hashing algorithms used to mine Bitcoin blocks, for example, take a "brute force" approach, which consists of systematically enumerating all possible candidates for the solution and checking whether each candidate satisfies the problem's statement before verifying a transaction. [43]

AI affords us the opportunity to move away from this and tackle tasks in a more intelligent and efficient way. Imagine a machine learning-based algorithm, which could practically polish its skills in 'real-time' if it were fed the appropriate training data. [43]

Creating Diverse Data Sets

Unlike artificial intelligence based-projects, blockchain technology creates decentralized, transparent networks that can be accessed by anyone, around the world in public blockchain networks situation. While blockchain technology is the ledger that powers cryptocurrencies, blockchain networks are now being applied to a number of industries to create decentralization. [48]

Data Protection

The progress of AI is completely dependent on the input of data—our data. Through data, AI receives information about the world and things happening on it. Basically, data feeds AI, and through it, AI will be able to continuously improve itself.

On the other side, blockchain is essentially a technology that allows for the encrypted storage of data on a distributed ledger. It allows for the creation of fully secured databases that can be looked into by parties who have been approved to do so. When combining blockchains with AI, we have a backup system for the sensitive and highly valuable personal data of individuals.

Medical or financial data are too sensitive to hand over to a single company and its algorithms. Storing this data on a blockchain, which can be accessed by an AI, but only with permission and once it has gone through the proper procedures, could give us the enormous advantages of personalized recommendations while safely storing our sensitive data. [44]

Data Monetization

Another disruptive innovation that could be possible by combining the two technologies is the monetization of data. Monetizing collected data is a huge revenue source for large companies, such as Facebook and Google.

Having others decide how data is being sold in order to create profits for businesses demonstrates that data is being weaponized against us. Blockchain allows us to cryptographically protect our data and have it used in the ways we see fit. This also lets us monetize data personally if we choose to, without having our personal information compromised. This is important to understand in order to combat biased algorithms and create diverse data sets in the future.

The same goes for AI programs that need our data. In order for AI algorithms to learn and develop, AI networks will be required to buy data directly from its creators, through data marketplaces. This will make the entire process a far fairer process than it currently is, without tech giants exploiting its users. [44]

Such a data marketplace will also open up AI for smaller companies. Developing and feeding AI is incredibly costly for companies that do not generate their own data. Through decentralized data marketplaces, they will be able to access otherwise too expensive and privately kept data.

11.2 Trusting AI Decision Making

As AI algorithms become smarter through learning, it will become increasingly difficult for data scientists to understand how these programs came to specific conclusions and decisions. This is because AI algorithms will be able to process incredibly large amounts of data and variables. However, we must continue to audit conclusions made by AI because we want to make sure they're still reflecting reality.

Through the use of Blockchain technology, there are immutable records of all the data, variables, and processes used by AIs for their decision-making processes. This makes it far easier to audit the entire process.

With the appropriate Blockchain programming, all steps from data entry to conclusions can be observed, and the observing party will be sure that this data has not been tampered with. It creates trust in the conclusions drawn by AI programs. This is a necessary step, as individuals and companies will not start using AI applications if they do not understand how they function, and on what information they base their decisions.

11.3 Conclusion

The combination of Blockchain technology and Artificial Intelligence is still a largely undiscovered area. Even though the convergence of the two technologies has received its fair share of scholarly attention, projects devoted to this groundbreaking combination are still scarce.

Putting the two technologies together has the potential to use data in ways never before thought possible. Data is the key ingredient for the development and enhancement of AI algorithms, and blockchain secures this data, allows us to audit all intermediary steps AI takes to draw conclusions from the data and allows individuals to monetize their produced data.

AI can be incredibly revolutionary, but it must be designed with utmost precautions—Blockchain can greatly assist in this. How the interplay between the two technologies will progress is anyone's guess. However, its potential for true disruption is clearly there and rapidly developing. [41]

12

Quantum Computing and Blockchain: Facts and Myths

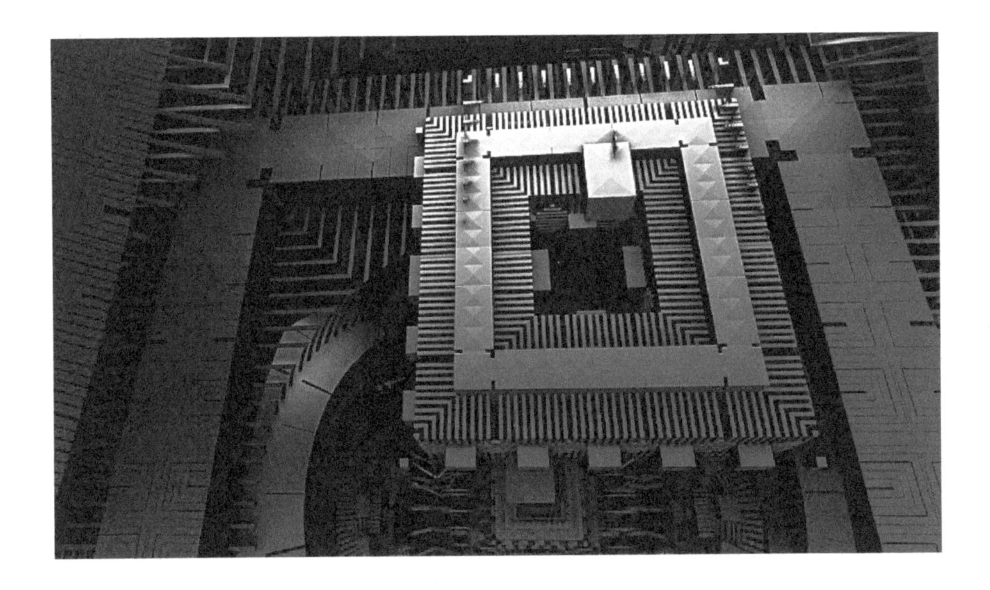

The biggest danger to Blockchain networks from quantum computing is its ability to break traditional encryption. [49]

Google sent shock waves around the internet when it was claimed, had built a quantum computer able to solve formerly impossible mathematical calculations–with some fearing crypto industry could be at risk [50]. Google states that its experiment is the first experimental challenge against the *extended Church-Turing thesis* — also known as computability thesis — which claims that traditional computers can effectively carry out any "reasonable" model of computation.

12.1 What is Quantum Computing?

Quantum computing is the area of study focused on developing computer technology based on the principles of quantum theory. The quantum computer, following the laws of quantum physics, would gain enormous processing power through the ability to be in multiple states, and to perform tasks using all possible permutations simultaneously. [51]

12.2 A Comparison of Classical and Quantum Computing

Classical computing relies, at its ultimate level, on principles expressed by Boolean algebra. Data must be processed in an exclusive binary state at any point in time or bits. While the time that each transistor or capacitor need be either in 0 or 1 before switching states is now measurable in billionths of a second, there is still a limit as to how quickly these devices can be made to switch state. As we progress to smaller and faster circuits, we begin to reach the physical limits of materials and the threshold for classical laws of physics to apply. Beyond this, the quantum world takes over. In a quantum computer, a number of elemental particles such as electrons or photons can be used with either their *charge* or *polarization* acting as a representation of 0 and/or 1. Each of these particles is known as a quantum bit, or *qubit*, the nature and behavior of these particles form the basis of quantum computing. [47]

12.3 Quantum Superposition and Entanglement

The two most relevant aspects of quantum physics are the principles of *superposition* and *entanglement* (Figure 12.1).

Superposition: Think of a qubit as an electron in a magnetic field. The electron's spin may be either in alignment with the field, which is known as a spin-up state or opposite to the field, which is known as a spin-down state. According to quantum law, the particle enters a superposition of states, in which it behaves as if it were in both states simultaneously. Each qubit utilized could take a superposition of both 0 and 1.

Entanglement: Particles that have interacted at some point retain a type of connection and can be entangled with each other in pairs, in a process known as *correlation.* Knowing the spin state of one entangled particle - up or down - allows one to know that the spin of its mate is in the opposite direction. Quantum entanglement allows qubits that are separated by incredible distances to interact with each other instantaneously (not limited to the speed of light). No matter how great the distance between the correlated particles, they will remain entangled as long as they are isolated. Taken together, quantum superposition and entanglement create an enormously enhanced computing power. Where a 2-bit register in an ordinary computer can store only one of four binary configurations (00, 01, 10, or 11) at any given time, a 2-qubit register in a quantum computer can store all four numbers *simultaneously*, because each qubit represents two values. If more qubits are added, the increased capacity is expanded exponentially. [47]

12.4 Difficulties with Quantum Computers

- Interference - During the computation phase of a quantum calculation, the slightest disturbance in a quantum system (say

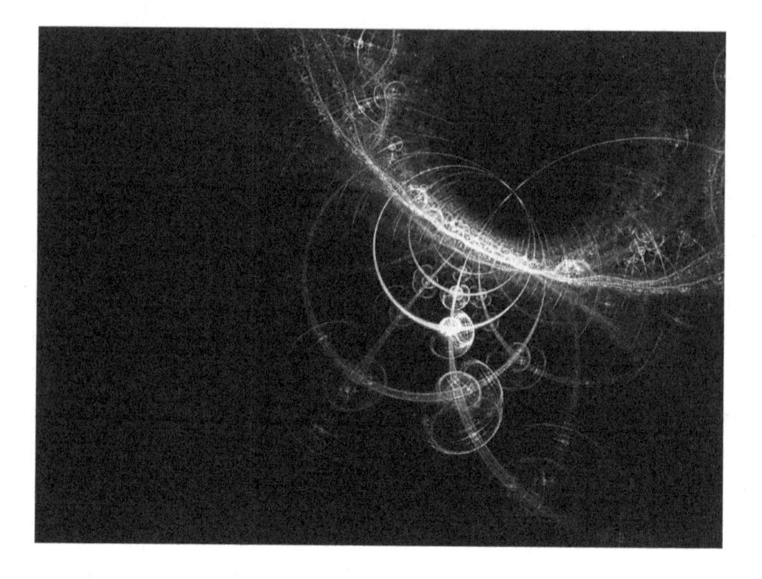

Figure 12.1: The principles of superposition and entanglement

a stray photon or wave of EM radiation) causes the quantum computation to collapse, a process known as *decoherence*. A quantum computer must be totally isolated from all external interference during the computation phase.

- Error correction - Given the nature of quantum computing, error correction is ultra-critical - even a single error in a calculation can cause the validity of the entire computation to collapse.
- Output observance - Closely related to the above two, retrieving output data after a quantum calculation is complete risks corrupting the data.

12.5 What is Quantum Supremacy?

According to the *Financial Times*, Google claims to have successfully built the world's most powerful quantum computer [46]. What that means, according to Google's researchers, is that calculations that normally take more than 10,000 years to perform, its computer was able to do in about *200 seconds*, and potentially mean Blockchain, and the encryption that underpins it could be broken.

Asymmetric cryptography used in crypto relies on key pairs, namely a private and public key. Public keys can be calculated from their private counterpart, but *not* the other way around. This is due to the impossibility of certain mathematical problems. Quantum computers are more efficient in accomplishing this by magnitudes, and if the calculation is done the other way then the whole scheme breaks. [45]

In order to have any effect on bitcoin or most other financial systems it would take at least about 1500 qubits and the system must allow for the entanglement of all of them.

Blockchain networks including Bitcoin's architecture relies on two algorithms: Elliptic Curve Digital Signature Algorithm (ECDSA) for digital signatures and SHA-256 as a hash function. A quantum computer could use Shor's algorithm [52] to get your private from your public key, but the most optimistic scientific estimates say that even if this were possible, it won't happen during this decade.

But that is not to say that there is no cause for alarm. While the native encryption algorithms used by Blockchain's applications are safe for now, the fact is that the rate of advancements in quantum technology is increasing, and that could, in time, pose a threat. "We expect their computational power will continue to grow at a double exponential rate," Google researchers.

12.6 Quantum Cryptography?

Quantum cryptography uses physics to develop a cryptosystem completely secure against being compromised without the knowledge of the sender or the receiver of the messages. The word *quantum* itself refers to the most fundamental behavior of the smallest particles of matter and energy.

Quantum cryptography is different from traditional cryptographic systems in that it relies more on *physics*, rather than mathematics, as a key aspect of its security model.

Essentially, quantum cryptography is based on the usage of individual particles/waves of light (photon) and their intrinsic quantum properties to develop an unbreakable cryptosystem (*because it is impossible to measure the quantum state of any system without disturbing that system*).

Quantum cryptography uses photons to transmit a key. Once the key is transmitted, coding and encoding using the normal secret-key method can take place. But how does a photon become a key? How do you attach information to a photon's spin?

This is where binary code comes into play. Each type of a photon's spin represents one piece of information — usually a 1 or a 0, for binary code. This code uses strings of 1s and 0s to create a coherent message. For example, 11100100110 could correspond with h-e-l-l-o. So a binary code can be assigned to each photon — for example, a photon that has a vertical spin (|) can be assigned a 1.

Regular, non-quantum encryption can work in a variety of ways but generally, a message is scrambled and can only be unscrambled using a secret key. The trick is to make sure that whomever you are trying to hide your communication from doesn't get their hands on your secret key. Cracking the private key in a modern cryptosystem would generally require figuring out the factors of a number that is the product of two insanely huge prime numbers.

The numbers are chosen to be so large that, with the given processing power of computers, it would take longer than the lifetime of the universe for an algorithm to factor their product.

Encryption techniques have their vulnerabilities. Certain products — called weak keys — happen to be easier to factor than others. Also, Moore's Law continually ups the processing power of our computers. Even more importantly, mathematicians are constantly developing new algorithms that allow for easier factorization.

Quantum cryptography avoids all these issues. Here, the key is encrypted into a series of photons that get passed between two parties trying to share secret information. The Heisenberg Uncertainty Principle dictates that an adversary cannot look at these photons without changing or destroying them.

13

Cryptocurrency: To Libra or not To Libra

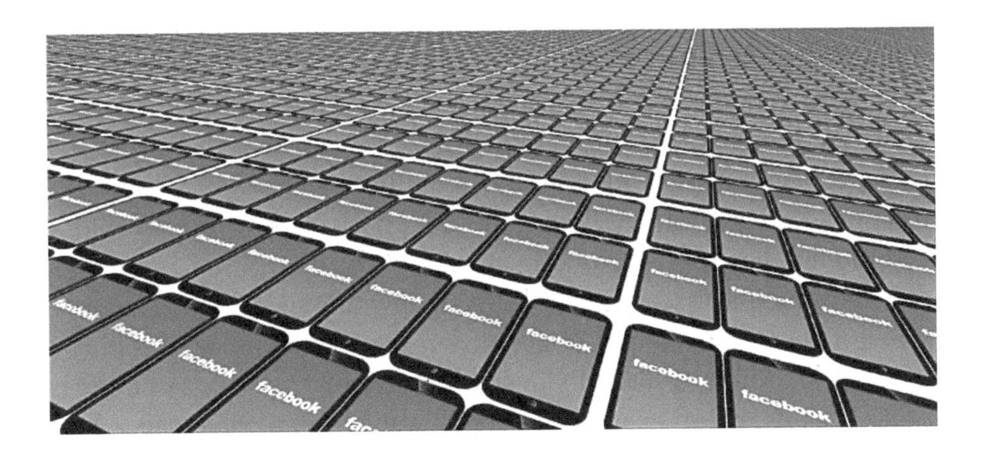

An interesting fact about Libra Facebook's native currency which was announced June 18th, 2019, it's inspired by three distinct elements: The *Roman weight* measurement system, the *astrological sign* for justice, and the *French term* for freedom. The culmination of these three elements embodies the essence of Libra, which aims to be a global cryptocurrency for everyone. The focus of Libra is to create a currency that empowers billions of people, allowing them to engage in friction-less financial transactions in a simple, secure, and cost-effective manner. [53]

But Facebook is launching *two* cryptocurrencies, not just Libra.

As part of June 18th big reveal of the social network's ambitious plan to create a global fiat-backed Blockchain currency, (Fiat currency is a government-issued currency like the dollar). Facebook said that in addition to Libra, the project will also issue a "*Libra investment token.*" Unlike Libra – a currency that will be broadly available to the public – the investment token is a security, according to Facebook. As such, the token will be sold to a much more exclusive audience: the founding corporate members of the project's governing consortium, known as the Libra Association, and accredited investors.

And while Libra will be backed by a basket of fiat currencies and government securities, interest earned on that collateral will go to holders of the investment tokens. Each of the 28 companies that Facebook recruited to run validating nodes as founding members of the consortium invested at least $10 million for the privilege. The investment token is what they received as a financial reward.

"Because the assets in the reserve are low risk and low yield, returns for early investors will only materialize if the network is successful and the reserve grows substantially in size," Facebook said in one of a series of documents that supplement the long-awaited Libra white paper. Further, the tokens will give holders proportional clout in the early governance of Libra. An investor who buys the tokens doesn't have to run a node, but unless they do, they don't get to vote as members. [54]

But every new trend, idea or concept in order to become mainstream must clear three obstacles namely: Technology, Business, and Society (Government and Customers).

The same applies to Libra, where, essentially, Facebook wants to make it as easy to move money around the world as it is to send a text message with lower fees, more accessibility, and close-to-instantaneous transfers worldwide. [49]

The company released a White Paper to explain the details. It doesn't see the cryptocurrency as an attempt to replace the current financial system, as is Bitcoin's aim. Rather, it's intended to extend a digital payment method to under-served populations that don't currently have easy access to traditional financial institutions. Worldwide, almost two billion adults "remain outside of the financial system with no access to a traditional bank, even though one billion have a mobile phone and nearly half a billion have internet access," reads the paper. Libra aims to fill the gap.

13.1 How will Libra work?

Libra will be managed by a Swiss-based nonprofit. It is currently backed by Facebook and less than two dozen Founding Member companies, including eBay, Uber, Lyft, Coinbase and venture capital firm Andreessen Horowitz. Unlike other cryptocurrencies, Libra will be backed by "real" government-backed assets from central banks to give it stability (Figure 13.1).

Facebook says Libra will be made available to Messenger and WhatsApp users, who can cash in their local currency to buy Libra. The currency will be held in a digital wallet called *Novi* and can be spent on products and services at participating merchants, just like any other currency.

To withdraw funds, users will be able to convert their digital currency into legal tender based on an exchange rate. It won't be so dis-

Figure 13.1: Facebook's Libra

similar to when you exchange U.S. dollars for euros during a European vacation, for example. For those worried about security, Libra payments will not be connected to a user's Facebook data and won't be used for ad targeting.

Libra will not be available until the end of 2020, so you cannot buy the currency today. Once it does become available, there should be several ways consumers can buy the currency, and you won't necessarily need to go through Facebook.

Transaction fees will likely be lower than those currently charged by traditional finance companies, which will primarily benefit merchants, but also people who, for example, routinely send money to family members abroad and are forced to rely on expensive wire transfer services.

13.2 How is this different from a credit card?

One of the purposes of Libra is to serve people who do not currently have access to traditional banking and financial tools. Currently, cryptocurrencies can be used like a credit card to buy goods online. But Libra will theoretically go beyond that. Consumers will be able to purchase the currency and use it at participating merchants (Figure 13.2).

"You have a balance of, say, $100, you go to a merchant, you scan your smartphone for a $10 purchase, the Libras are taken out of your account and held by the merchant," Transaction fees will also be lower than they are for traditional forms of payment. Novi is similar, then, to a payment network like PayPal, but uses the cryptocurrency Libra rather than a fiat currency, like the U.S. dollar for transactions. [55] Libra will be a stable digital currency, which will be fully backed by real assets stored in the Libra reserve. Stablecoins are cryptocurrency that is stable in value, usually pegged from a real-world currency (such as US$) or a commodity (such as gold) that are stable in nature.

The Libra reserve is created through funds originating from both investors in the separate Investment Token, and users of Libra. This means that you can invest in the project through an Investment Token

Figure 13.2: How Libra works

that could potentially pay out dividends in the future or if you are keen on getting your hands-on Libra coin itself, you have to convert your local currency into Libra.

Essentially, Libra is only created when there is more fiat with which it is either exchanged or backed up. The Libra reserve will then be invested in low-risk assets that will yield interest over time, which will then be used to cover operational costs, support low transaction fees, and pay dividends to Investment Token investors who helped jump-start the ecosystem. The stability of Libra will, therefore, be supported by a global basket of fiat currencies and low-risk assets; likewise, you can convert your Libra at any time according to the prevailing exchange rate to your local currency.

Facebook will relinquish control over Libra, instead of conferring control to Libra Association, a Geneva-based non-profit organization with a long list of prominent founding members, including, Lyft, Spotify, Uber, and Coinbase. Currently, there are 21 members, each of whom are required to invest $10 million into the development of Libra. The Libra foundation aims to accumulate a total of 100 partners with a reserve fund of $1 billion, which is going to be used to manage Libra's price stability. All members will also be granted a single vote for the governance of Libra, with each entity serving as nodes in the Libra network.

The association will also spearhead Libra's native open-source technology, by promoting its developer platform, which is fueled by its own programming language. Given the wide reach of the cumulative networks of all members in the association, it is not hard to imagine that there will be a colossal base of ready users for Libra, something which would have been incomprehensible with any other past crypto-currency projects, particularly on this scale.

13.3 Technology

Libra will be built on the Libra Blockchain, a natively developed open-sourced Blockchain that uses a Byzantine Fault Tolerant (BFT) consensus approach called LibraBFT Consensus Protocol (voting based protocol used in Hyperledger networks). However, the Libra Blockchain will initially be a permissioned (closed) Blockchain. This means that access to the network is limited to a handful of selected and pre-approved entities who will become nodes in the system.

Libra will gradually transit into a permissionless (public) network (similar to Bitcoin and Ethereum networks) within five years of the public launch of Libra blockchain and ecosystem. The rationale behind this is that a permissionless network has limitations in terms of speed and scalability, and in order to deliver a scalable, secure, and stable solution globally across billions of people and transactions, it needs to be a permissioned system at first. That said, Libra Blockchain will be open in the sense that anyone can use the network and even build applications on top of the Blockchain. [56]

13.4 The Libra Blockchain

With 5KB transactions, 1000 verifications per second verifications on commodity CPUs, and up to 4 billion accounts, the Libra Blockchain should be able to operate at 1000 tps (transactions per second) if nodes used at least 40Mbps connections and 16TB SSD hard drives. Transactions on Libra cannot be reversed. If an attack compromises *over one-third* of the validator nodes causing a fork in the Blockchain, the Libra Association says it will temporarily halt transactions, figure out the extent of the damage, and recommend software updates to resolve the fork. [52]

Libra Blockchain will facilitate *smart contract* functionality, using a relatively new language called 'Move'. Smart contracts are pre-programmed contracts that are self-executable, thereby allowing for the automation of contracts without the need of any third parties or intermediaries. Move is a simple but powerful language, and is relatively suitable as a "first programming language". More importantly, Move is designed with a key focus on security and safety, since leveraging a simpler language facilitates easier code writing and execution, and reduces the risk of unintended bugs or security flaws (Figure 13.3).

The first application to be built to support a cryptocurrency must be a wallet. This is exactly the case with Libra, in which Novi, a cryptocurrency wallet, will facilitate the storage and exchange of Libra coins. Novi will be the first application to be built on the Libra Blockchain.

Novi is also the name of the company which will develop the wallet and is, in fact, a subsidiary of Facebook, built to ensure separation

Figure 13.3: Libra and Blockchain

between financial and social data and to build and operate services on its behalf on top of the Libra Blockchain.

Novi will be available as a mobile application and will also be integrated with Facebook's Messenger application and WhatsApp, allowing users to convert fiat currency into Libra in their wallets and thereafter send, receive, and pay for stuff using Libra. [52]

13.5 Business

According to Facebook, almost half of adults in the world do not have an active bank account, with the figures worse in developing countries and even worse for women. Approximately 70% of small businesses in developing countries lack access to credit, and $25 billion is lost by migrants annually through remittance fees. [57]

Facebook has more than 1.5 billion users on both WhatsApp and Messenger yet makes almost no money from the messaging services. When Facebook revealed its Libra plans, the company also said it would soon put new digital wallets inside these apps so users can easily use the cryptocurrency to send money to friends and businesses anywhere

Figure 13.4: Libra Business Reach

in the world. If the plan works, WhatsApp and Messenger will become new payments and commerce hubs that take small-but-profitable cuts from billions of transactions (Figure 13.4).

Facebook has a checkered record in payments. But China's WeChat and QQ show what's possible when messaging apps cleverly fold payments and other services into the mix. WeChat and QQ make money by facilitating payments between users and merchants, distributing mobile games, and selling digital goods, such as stickers and avatars. The services have turned owner Tencent Holdings Ltd. into the most valuable publicly traded company in China.

Facebook's crypto push could facilitate similar offerings in payments, shopping, apps, and gaming, while tapping into the company's huge user base in Asia, where it has nearly four times as many monthly active users as it does in North America, according to RBC Capital Markets.

For now, Facebook and its new subsidiary Novi, which is building the digital wallets, are framing the new currency as a way for individuals to send money to each other across borders. David Marcus, who is leading Facebook's Libra efforts, said that the company does not plan to take a fee when people send money to friends, and will likely charge "tiny transaction fees" for payments to businesses.

If people do start stuffing their new digital wallets with Libra, it might not take years for Facebook to turn that activity into revenue. Marcus believes the new wallets could have a more immediate financial impact on a business line Facebook knows well: Targeted advertising. If users have Libra on hand as they scroll through Facebook's News Feed, when they click on an ad it will be easier to buy something. That would make Facebook ads more appealing to marketers.

"If there is more commerce happening on the platform, then small businesses will end up spending more and advertising will be more effective for them," Marcus said. [58]

To generate higher levels of adoption among users, Libra has developed an incentive program to encourage more developers to create applications on Libra Blockchain, and more merchants to accept Libra as a payment currency. Node operators, who represent the founding members of the Libra association, will be rewarded with Libra coins for getting users to sign up and use Libra. Businesses that attract users towards Novi will also be rewarded with incentives that they can pass on, in part or in their entirety, to users in the form of discounts or free Libra tokens for their purchases. Merchants in the network are also incentivized by receiving a percentage of the transaction value back for each transaction that is processed on the platform.

The incentive programs are targeted towards the entire Libra network, ensuring that a holistic approach is undertaken to foster adoption in the usage of Libra.

13.5 Society

13.5.1 Government

The company's crypto plans are already under fire from regulators in Washington and Europe who don't like the idea of Facebook dipping its toe in yet another aspect of people's personal lives. And gaining consumer trust after years of privacy mishaps may be harder than Facebook expects. A letter from the U.S. House Representatives Committee on Financial Services to Facebook sent on July 2, Facebook to officially put the project on hold.

Today, cryptocurrencies are backed solely by the willingness of users to accept them, not because they have any intrinsic value or are backed by any government. This makes such currencies unstable. Libra, however, will be backed by reserves: If a user buys a dollar of Libra, that dollar will presumably be held in reserve somewhere, ready to be honored when someone sells that Libra. Moreover, while most cryptocurrencies are hard to use, Libra promises to be user-friendly and embedded into Facebook and WhatsApp.

There are *four core problems* with Facebook's new currency. [59]

The first, and perhaps the simplest, is that organizing a payments system is a complicated and difficult task, one that requires an enormous investment in compliance systems. Banks pay attention to details, complying with regulations to prevent money-laundering, terrorist financing, tax avoidance, and counterfeiting. Recreating such a complex system is not a project that an institution with the level of privacy and technical problems like Facebook should be leading. [55]

The second problem is that, since the Civil War, the United States has had a general prohibition on the intersection between banking and commerce. Such a barrier has been reinforced many times, such as in 1956 with the Bank Holding Company Act and in 1970 with an amendment to that law during the conglomerate craze. Both times, Congress blocked banks from going into nonbanking businesses through holding companies, because Americans historically did not want banks competing with their own customers. Banking and payments is a special business, where a bank gets access to intimate business secrets of its customers. [55]

Imagine Facebook's subsidiary Novi knowing your account balance and your spending, and offering to sell a retailer an algorithm that will maximize the price for what you can afford to pay for a product. Imagine this cartel having this kind of financial visibility into not only many consumers, but into businesses across the economy. Such conflicts of interest are why payments and banking are separated from the rest of the economy in the United States. [55]

It is also possible that insiders belonging to the Libra cartel could exploit their acccss to information, business relationships or technology to give themselves advantages. There are many ways a new currency system could advantage large businesses over everyone else, especially when the large ones are sitting on the board of governors for the payments system. For instance, one of the incentives is to get people to use the currency is discounts on Uber rides; if this happens, Facebook would be giving an advantage to Uber instead of other ride-sharing businesses. [55]

The third problem is that the Libra system introduces systemic risk to our economy. The Libra currency is backed, presumably, by bonds and financial assets held in reserve at the Libra Reserve. But what happens if there is a theft or penetration of the system? What happens if all users want to sell their Libra currency at once, causing the Libra Reserve to hold a fire sale of assets? If the Libra system becomes intertwined in our global economy in the way Facebook hopes, we would need to consider a public bailout of a privately managed system. We should not be setting up a private international payments network that would need to be backed by taxpayers because it is too big to fail. [55]

The fourth problem is that of national security and sovereignty. Enabling an open flow of money across all borders is a political choice best made by governments. And openness is not always good. For instance, most nations, especially the United States, use economic sanctions to bar individuals, countries or companies from using our financial system in ways that harm our interests. Sanctions enforcement flows through the banking system—if you cannot bank in dollars, you cannot use dollars. With the success of a private parallel currency, government sanctions could lose their bite. A permissionless (public) currency system based on a consensus of large private actors across open protocols sounds nice, but it is not democracy. Today, American bank regulators and central bankers are hired and fired by publicly elected leaders. Libra payments regulators would be hired and fired by a self-selected council of corporations. There are ways to characterize such a system, but democracy is not one of them. [55]

The social media giant says it will accept the cryptocurrency anywhere where it takes payments and at least for now won't rule out allowing it to be used to buy political ads. [60]

The Libra Association, a not-for-profit organization based in Switzerland, will have several layers of governance, the most powerful of which is a council, on which each member organization will have a representative.

"The council delegates many of its executive powers to the association's management but retains authority to override delegated decisions and keep key decisions to itself, with the most important ones requiring a greater than two-thirds supermajority," according to another supplementary document released by Facebook. As mentioned, to become a member, the initial investors must put in at least $10 million. In addition, a business must meet at least one of several elite criteria, such as being on a list like the Fortune 500.

For every $10 million invested, a member gets one vote, subject to a cap of 1 percent of total votes, in order to prevent the concentration of power in any single entity. However, the financial reward remains proportional to the amount invested no matter how much.

The council will be responsible for standard governance matters, such as appointing an executive team for the association, led by a managing director, and a board of directors to oversee them; setting the top executive's compensation, and managing the currency's underlying reserves.

But the body will also have final say over technical questions, such as activating new features to the protocol and resolving situations "where compromised validator nodes have resulted in many signed versions of the Libra Blockchain," according to the document.

While Facebook's newly created Novi subsidiary will be a consortium member with a council seat, the social network stressed it won't be in charge for long. "Once the Libra network launches, Facebook, and its affiliates, will have the same commitments, privileges, and financial obligations as any other Founding Member," the company said. "As one member among many, Facebook's role in the governance of the association will be equal to that of its peers."

The exact components of the basket of assets securing Libra are to be determined. But broadly, it will be "structured with capital preservation and liquidity in mind," according to the social media giant. Importantly, while the coin has been described in early press coverage as a stablecoin, Facebook noted that "from the point of view of any specific currency, there will be fluctuations in the value of Libra."

"The makeup of the reserve is designed to mitigate the likelihood and severity of these fluctuations, particularly in the negative direction (i.e., even in economic crises)." In this way, Libra will function more like a currency board such as Hong Kong's rather than a central bank. The collateral will consist of "bank deposits and government securities in currencies from stable and reputable central banks," according to Facebook. The latter will be limited to "debt from stable governments that are unlikely to experience high inflation."

To make sure it can easily raise cash by selling this paper, it will all be "short-dated securities issued by these governments that are all traded in liquid markets." While the composition of the basket may change over time, Facebook said, the currency will always be fully backed, discouraging "runs on the bank" that can happen with fractional reserve institutions. [50]

To comply with anti-money-laundering regulations that require traceability of funds, transactions on the Libra Blockchain will be *unencrypted*, "like many other Blockchain, so it is possible for third parties to do analysis to detect and penalize fraud," Facebook said. In other words, it appears that there will be no use of cryptographic mechanisms such as zero-knowledge proofs, used to obscure transaction details in privacy-focused coins such as zcash.

If that raises privacy concerns (particularly given Facebook's own reputation with user data), the company is offering similar assurances to those Satoshi Nakamoto gave in the 2008 bitcoin white paper. [50]

13.5.2 Customers

It is important to realize that Facebook is actually launching two cryptocurrencies: the one everyone's talking about (Libra) and the one available only to Facebook and its corporate partners (the Libra investment token).

The former will be backed by a basket of fiat currencies and cash equivalents, which means that for every dollar of Libra in existence, there will be (in theory) a "dollar" worth of real-world assets which that token may be exchanged for under certain conditions.

As a normal user, you would get $100 worth of Libra by spending $100. Your Libra can (again, in theory) be used across a variety of platforms or sent to an approved friend.

The Libra Association (a Swiss not-for-profit) puts your $100 into a variety f low-risk, short-term investments like U.S. Treasury bills. Those funds are controlled and spent by the Libra Association. According to the white paper, funds are used first to fund the operation of the network with the remainder being divided among the Libra Investment Token holders according to their holdings, with policies determined by the association.

The association itself is made up of holders of the Libra investment token who invested a minimum of $10 million, as well as "special impact groups" selected by the association to have a vote but who do not have to buy the investment token.

From the white paper:

> "*How will the reserve be invested? Users of Libra do not receive a return from the reserve. The reserve will be invested in low-risk assets that will yield interest over time. The revenue from this interest will first go to support the operating expenses of the association — to fund investments in the growth and development of the ecosystem, grants to nonprofit and multilateral organizations, engineering research, etc. Once that is covered, part of the remaining returns will go to pay dividends to early investors in the Libra Investment Token for their initial contributions. Because the assets in the reserve are low risk and low yield, returns for early investors will only materialize if the network is successful and the reserve grows substantially in size.*" [61]

Early investors are primarily large technology and VC companies, for whom $10 million is not actually a huge investment. The big numbers come into play when you look at what success, big or small, would look like for the investors, at which point suddenly the project makes sense. [57] Consumers, who will decide ultimately whether or not Libra is a flop, there was only a slightly underwhelming hint of what it might actually be used for: A picture of someone sending money to someone else via a smartphone.

Even setting aside the various risks thrown up by the Libra white paper (financial stability, user privacy, and whether it could cope with hundreds of millions of daily transactions), you have to ask why it might be a compelling product. The service described by Facebook, namely sending money "as you might send a text message," is already offered by plenty of other companies such as Alphabet Inc.'s Google, Apple pay, PayPal Holdings Inc.'s Venmo and Circle, a peer-to-peer payments provider that lets you transfer traditional fiat currencies.

Indeed, Facebook itself lets you send cash through its Messaging app. The company even had its own virtual currency before, called Credits, for the purchasing of content from within apps. It did not take off.

Facebook plans to lead the Libra consortium for the near future, and it will be at least five years before the Blockchain technology that supports the tokens is completely decentralized. The ultimate dream of any crypto project worth its salt is that the digital currency does not rely on a single point of control. And what about Facebook's targeting of the "unbanked," or those in the developing world struggling with volatile currencies? Bitcoin and its ilk promised to address the same problems, and have failed completely to help anyone other than speculators and criminals.

Facebook's own patchy record on international payments should give pause too. WhatsApp Pay has struggled to gain regulatory acceptance in India, the world's top remittance market because its data storage practices did not meet national standards. Libra will have to answer a lot of similar questions about its financial structure and treatment of customer information.

Facebook has been on a mission over the past year to recapture the trust of its users. Libra certainly demands a lot of faith. [62]

Another big question surrounding Libra is what kinds of consumer protections — if any — Facebook and its partners will build into the system. With bitcoin and other cryptocurrencies, there are often few protections.

Stories abound of people losing thousands or millions of dollars or more because they cannot remember the passcode to their cryptocurrency wallet, or a hacker broke into their wallet or into the cryptocurrency exchange where it was stored. And typically, any transactions conducted using such digital currency are final once they happen.

They cannot be reversed because one party made a mistake or a customer did not get what he or she ordered. Customers using the traditional banking system have many more protections. In the US, federal banking insurance covers deposits. Consumers generally have the right to contest charges or have transactions reversed if they did not receive what they have ordered or were defrauded. And they generally would not be held responsible if someone steals their credit card and uses it to make a bunch of purchases.

It's unclear what model Facebook and its partners will follow with Libra. But it is likely that many customers and, perhaps, regulators are going to expect the company to offer similar protections that banks and credit card issuers offer. [63]

13.6 Conclusion

Libra is still in the early stages of development with lots of things left to do, with a targeted launch by end of 2020. However, Libra is perhaps the most ambitious and hyped cryptocurrency in existence, drawing on the stature of Facebook as a unicorn, as well as the partnerships that have been developed by the Libra association. With plans to create a fully-functioning Blockchain that is open-source, and facilitates smart contract technology using a native programming language, Libra seems to be moving in fundamentally the right direction.

More importantly, the adoption of cryptocurrency and Blockchain technology by a giant technology firm has set the tone for mainstream adoption of this nascent technology, and the fact that a prominent list of institutions, all of which were previously openly reluctant to embrace cryptocurrencies, serve as members of the Libra association, is a huge testament to the changing tides of acceptance towards distributed ledger technologies. Saying that this is a huge deal is perhaps an understatement, but it is definitely a well-deserved victory for the industry and technology. [49]

14

Future Trends of Blockchain

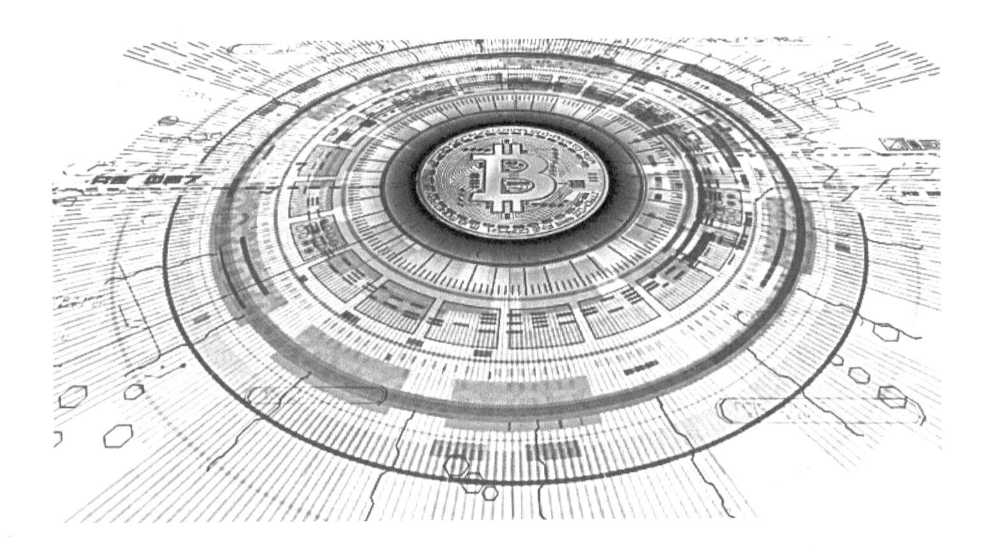

It's clear that Blockchain will revolutionize operations and processes in many industries and governments agencies if adopted, but its adoption requires time and efforts, in addition, Blockchain technology will stimulate people to acquire new skills, and traditional business will have to completely reconsider their processes to harvest the maximum benefits from using this promising technology.

14.1. A Reality check for Blockchain

Blockchain is suffering from the same type of over-hype that virtual reality is. For years, people in various sectors have been hearing about Blockchain. It's been portrayed as a true game-changer. The problem is that so many people still aren't seeing real-world benefits. To them, like virtual reality, it remains a nifty technology without a practical application they can really wrap their heads around.

In 2018, we saw an increase in funding for Blockchain startups. However, like any new technology, Blockchain is still immature in its implementation; as a result, many Blockchain startups are expected to be just a waste of time and money. False starts in Blockchain deployment will lead organizations to failed innovations, rash decisions, and even complete refusal of this innovative technology.

Undoubtedly, Blockchain technology in the future will affect every aspect of businesses, but this is a gradual process that requires time and patience. Gartner predicts that most traditional businesses will keep an eye on Blockchain technology, but won't plan any actions, waiting for more examples of the best applications of Blockchain technology.

The reason for this is that traditional enterprises require more transformation for Blockchain deployment than newly-appeared businesses. According to Gartner, only 10% of traditional companies will achieve any radical transformation with Blockchain technologies by 2023. [64]

This is not to say that Blockchain doesn't have amazing potential. It certainly does prove that the gap between Blockchain hype and application is a problem. If businesses stop exploring their potential, it won't matter how useful the technology is. But this may be changing as well. [65]

14.2. The 'Emerging Disruptor' startups fueled by Blockchain

The Blockchain 'emerging disruptor' companies are fast-growth startups that have found themselves in the position to be able to disrupt other businesses in their sector. They often have the benefit of being well-funded and headed by executives who are experienced and well-connected in their industries.

These are the businesses that are often able to apply Blockchain technologies in ways that are truly a part of their business model, as opposed to supplementing it. Competing with big names like Amazon, Google, Facebook, Apple, and Microsoft. [61]

For example, Blockchain could be useful for content streaming companies like Netflix because it could be used to store data more securely and to pave the way for interoperability. If nothing else, it could provide something resembling an API, allowing third-parties to read and write data to the Blockchain.

But there could be a more practical use for Blockchain in the streaming industry and in other industries that require large amounts of processing power. In the same way that mining for bitcoin just taps into dormant computing power from across a network of machines, streaming companies could witness huge decreases in their operating costs by spreading the load across unused machines via the Blockchain. [66]

14.3. Blockchain and Cybersecurity

The global figure for cyber breaches had been put at around $200 billion annually [67]. Because Blockchain was created as a means to ensure the security of transactions, it shouldn't come as a big surprise that this is the niche where much of the innovation still occurs. Blockchain is playing a huge role in cybersecurity especially. [61] With the growing prevalence of data breaches and the massively interconnected world we live in, new ways to verify identity and protect privacy will be game-changers. Blockchain is a natural for this role because the whole point of it is to provide robust, incorruptible — yet encrypted — record-keeping that anyone can easily verify.

An example of such an application, Blockchain can be used for shopping security, whether online or in person. Blockchain in this space can create a "universal shopper profile" that is undergirded by Blockchain. Unlike most systems these days, in which your purchase histories are stored and carefully scrutinized and shared by big names such as Google, Blockchain restricts the information collection and sharing to only those entities that you (the consumer) grant it to share, and give consumers incentive to see ads by tokenizing the process and give rewards. [68]

For a Blockchain system to be penetrated, the attacker must intrude into every system on the network to manipulate the data that is stored on the network. The number of systems stored on every network can be in millions. Since domain editing rights are only given to those who require them, the hacker won't get the right to edit and manipulate the data even after hacking a million systems. Since such manipulation of data on the network has never taken place on the Blockchain, it is not an easy task for any attacker.

While we store our data on a Blockchain system, the threat of a possible hack gets eliminated. Every time our data is stored or inserted into Blockchain ledgers; a new block is created. This block further stores a key that is cryptographically created. This key becomes the unlocking key for the next record that is to be stored onto the ledger. In this manner, the data is extremely secure.

Furthermore, the hashing feature of Blockchain technology is one of its underlying qualities that makes it such a prominent technology. Using cryptography and the hashing algorithm, Blockchain technology converts the data stored in our ledgers. This hash encrypts the data and stores it in such a language that the data can only be decrypted using keys stored in the systems. Other than cybersecurity, Blockchain has many applications in several fields that help in maintaining and securing data. The fields where this technology is already showing its ability are finance, supply chain management, and Blockchain-enabled smart contracts. [63]

14.4. Internet of Things (IoT) meets Blockchain

It's no secret that the internet of things is coming to connect our devices and to make it easier than ever for us to create and store data about ourselves. This applies to everything from wearable devices to home hubs, connected fridges and any other type of internet-connected device that you can imagine.

But all of these internet-connected devices will need some sort of security system that ties them together and that makes their data secure. That could be where the Blockchain comes in, but only if different manufacturers can agree to come together and agree on the specifications of the Blockchain that's required. [62]

The International Data Corporation (IDC) reports that many #IoT companies are considering the implementation of Blockchain technology in their solutions. Therefore, IDC expects that nearly 20 percent of IoT deployments will enable Blockchain services in two years.

The reason for this is that Blockchain technology can provide a secure and scalable framework for communication between IoT devices. While modern security protocols already appeared to be vulnerable when implemented to IoT devices, Blockchain has already approved its high resistance to cyber-attacks. [60]

Besides, Blockchain will allow smart devices to make automated micro-transactions. Due to its distributed nature, Blockchain will conduct transactions faster and cheaper. To enable transferring money or data, IoT devices will leverage smart contracts which will be considered as the agreement between the two parties. [60]

Since the beginning of 2018, analysts have been predicting that IoT DApps might just become the next key development in Blockchain application development. In 2019, 20% of all IoT deployments have at least basic levels of Blockchain services enabled.

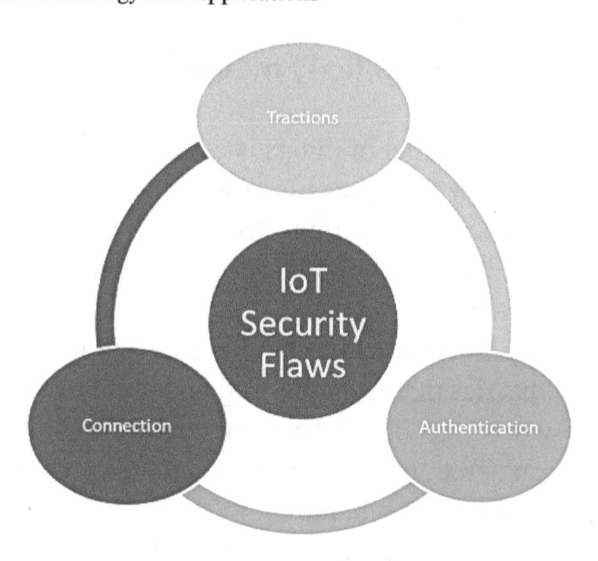

Figure 14.1: IoT Security Flaws

IoT security flaws generally revolve around three major events: authentication, connection, and tractions (Figure 14.1). Thanks to these vulnerabilities, already hackers have managed to take control of implanted cardiac devices, disable cars remotely, and launch the largest DDos attack to date.

Given that security is one of the main challenges of IoT, as well as, data integrity, it goes without saying that Blockchain could potentially revolutionize this sector. Blockchain technology makes it possible to stabilize complex IoT systems. It also eliminates the risk of single points of failure for an IoT network as a result of malicious attacks. [69]

14.5. Increased use of smart contracts

Smart contracts are one of the most interesting aspects of Blockchain technology because they have the potential to bypass third parties and to create airtight agreements that must be honored. This has plenty of practical applications in all sorts of industries, from finance and real estate to logistics and recruitment. Any industry that relies on agreements to function can get a lot out of smart contracts.

The idea behind these contracts is that they offer increased transparency and security while simultaneously speeding up the whole process. Contracts could be signed and verified in real-time in a secure environment, and that can make all of the difference when it comes to getting things done and reacting quickly to changes in the market [62].

You should also keep in mind that smart contracts are decentralized and aren't regulated by any authority. But what should parties do in case of any disagreement? Participants of smart contracts usually agree to be bound by regulations, but what if a dispute appears between parties from different countries. Now, it remains to be unclear how contractual disputes should be settled. Thus, the rule of law should be enforced into smart contracts in the near future for resolving any disputes between the parties. [60]

14.6. Increased regulations

As Blockchain becomes more widely used in all sorts of different industries, it'll also start to receive more attention from regulators and lawmakers, especially if governments can find ways to use Blockchain technology on a large scale. And let's not forget that cryptocurrency is technically classed as a property and not a currency when it comes to taxation, at least in the United States.

Whenever a new innovation like Blockchain comes along and starts to create large sums of money for those who are able to take advantage of it, it tends to receive intense scrutiny from people in power. But that is not necessarily bad news for Blockchain, because additional interest in it from powerful organizations may help to increase consumer trust whilst providing the framework needed for further growth. [62]

14.7. Financial institutions will lead to Blockchain evolution and revolution

Unlike other traditional businesses, the banking and finance industries do not need to introduce radical transformation to their processes for adopting Blockchain technology. After it was successfully applied for the cryptocurrency, financial institutions begin seriously considering Blockchain adoption for traditional banking operations.

In a recent PwC report, 77 percent of financial institutions are expected to adopt Blockchain technology as part of an in-production system or process by end of 2020.

Though the concept of Blockchain is simple, it will bring considerable savings for banks. Blockchain technology will allow banks to reduce excessive bureaucracy, conduct faster transactions at lower costs, and improve its secrecy. One of the Blockchain predictions made by Gartner is that the banking industry will derive 1 billion dollars of business value from the use of Blockchain-based cryptocurrencies by end of 2020.

Moreover, Blockchain can be used for launching new cryptocurrencies that will be regulated or influenced by monetary policy. In this way, banks want to reduce the competitive advantage of standalone cryptocurrencies and achieve greater control over their monetary policy. [60]

14.8. National cryptocurrencies will appear

It's inevitable that governments will have to recognize the benefits of Blockchain-derived currencies. At the rise of Bitcoin, governments expressed their skepticism regarding the particular application of cryptocurrencies. However, they had to worry when Bitcoin became a tradeable currency that could not be controlled by any government.

Although some countries like China still ban Bitcoin exchanges, we should expect that governments will finally accept the Blockchain-based currency in 2019 because of its potential advantages for public and potential services. By end of 2022, Gartner predicts that at least five countries will issue a national cryptocurrency. [60]

14.9. Blockchain integration into government agencies

The idea of the distributed ledger is also very attractive to government authorities that have to administrate very large quantities of data. Currently, each agency has its separate database, so they have to constantly require information about residents from each other. However, the implementation of Blockchain technologies for effective data management will improve the functioning of such agencies.

According to Gartner, by 2022, more than a billion people will have some data about them stored on a Blockchain, but they may not be aware of it. [60]

14.10. Blockchain experts will be in high demand

Despite Blockchain is on the top of its popularity, the job market experiences a lack of Blockchain experts. Upwork, an online freelancing database, has recently reported a fast-increasing demand in people with "Blockchain" skills. While the technology is new, there are a limited number of Blockchain engineers. [60]

14.11. Blockchain and Artificial Intelligence (AI)

One of Blockchain's most promising use cases lies in the fact that it has to potential to facilitate certain parts of an AI implementation. In order for AI to function, machines require access to big data. Up to now the processing of big data has not been economically viable. However, with the support of the Blockchain, this may all change.

Blockchain can provide the data authentication on which AI models depend since the data stored on the ledger cannot be changed and is available publicly. That makes data stored in a Blockchain more relevant than data that is delivered on unproven platforms that have embedded errors. [65]

The quest for artificial intelligence has been a long-standing one. Ever since the emergence of computers, scientists have been looking for ways to develop thinking machines. AI is basically an algorithm that allows machines to exhibit functions that they were not programmed for. The most complex devices on the planet still only work within the limits of their programming algorithms. When successfully implemented, AI will birth machines that can quote "learn how to learn."

Just like in the case of the IoT, the Blockchain has been identified as having the potential to facilitate certain aspects of the AI implementation. In order to function to its fullest capacity, machines capable of learning require access to "big data." The majority of the big data available for mainstream use is reserved for analytics. Exchange of big data hasn't been economically feasible but with the aid of the Blockchain, this could all change.

Blockchains can provide a secure environment for big data owners to connect with AI developers. By so doing, complex machine learning algorithms can be developed to help smart devices take advantage of the data available to them in order to achieve artificial sentience. [70]

14.12. New Blockchain Platforms with Better Processing Power and UX

Ethereum is a well-known platform for Blockchain technology but it has a major fundamental flaw; its highly inefficient usage of the Blockchain's processing power, other Blockchain platforms are emerging to fill the gap. Instead of Ethereum, DApp developers are turning to other Blockchains, such as Hyperledger.

Hyperledger offers a major advantage over Ethereum because it allows developers to create DApps with private Blockchains, as well as, permissioned Blockchains. Hyperledger offers low node-scalability which enables high-performance scalability. With Hyperledger, nodes can also assume different roles and tasks in order to reach a consensus that enables fine-grained control over consensus.

While Blockchain projects have mostly been focused on taking advantage of the versatility of Blockchain technology, usability has been severely overlooked. In 2019, you can expect to see new projects that aim to make things easier for everyone, for end-users, as well as, developers.

New platforms are making things easier for developers with functional programming languages and easy-to-deploy and customizable Blockchains. On the user end, the end goal is for users to not even know that they are using Blockchain technology. For example, Blockchain developers are building on platforms, which don't require users to pay fees. [65]

14.13. Blockchain as a Service (BaaS) By Big Tech Companies

One of the promising blockchain trends in 2020 is BaaS, short for Blockchain As A Service. It is a new blockchain trend that is currently integrated with a number of startups as well as enterprises. BaaS is a cloud-based service that enables users to develop their own digital products by working with blockchain. These digital products may be smart contracts, decentralized applications (DApps), or even other services that can work without any setup requirements of the complete blockchain-based infrastructure.

Some of the companies developing a blockchain that provide BaaS service are Microsoft and Amazon, consequently shaping the future of blockchain applications.

14.14. Federated Blockchain Moves to The Center Stage

Blockchain networks can be classified as Private, Public, Federated or Hybrid. The term Federated Blockchain can be referred to as one of the best blockchain latest trends in the industry. It is merely an upgraded form of the basic blockchain model, which makes it more ideal for many specific use cases.

In this type of blockchain, instead of one organization, multiple authorities can control the pre-selected nodes of blockchain. Now, this selected group of various nodes will validate the block so that the transactions can be processed further. In 2020, there will be a rise in the usage of federated blockchain as it provides private blockchain networks, a more customizable outlook.

14.15. Stablecoins Will Be More Visible

Using Bitcoin as an example of cryptocurrencies its highly volatile in nature. To avoid that volatility stablecoin came to the picture strongly with stable value associate with each coin. As of now, stablecoins are in their initial phase and it is predicted that 2020 will be the year when Blockchain stablecoins will achieve their all-time high.

One driving force for using stablecoin is the introduction of Facebook's cryptocurrency "Libra" in 2020 even with all the challenges facing this new cryptocurrency proposed by Facebook and the shrinking circle of partners in libra.org.

14.16 Final Thoughts

There's nothing but opportunity ahead for businesses who want to utilize Blockchain. However, most Blockchain application development trends require more than just developers. You'll also need to make changes to your workforce, as well as, your overall business strategy in order to effectively leverage the benefits of Blockchain technology.

Special Topics in Blockchain

15

Blockchain Technology and COVID-19

The COVID-19 coronavirus has impacted countries, communities and individuals in countless ways, from school closures to health-care insurance issues not to undermined loss of lives. As governments scramble to address these problems, different solutions based on blockchain technologies have sprung up to help deal with the worldwide health crisis. Figure 15.1 explains the source of the name COVID-19. [71]

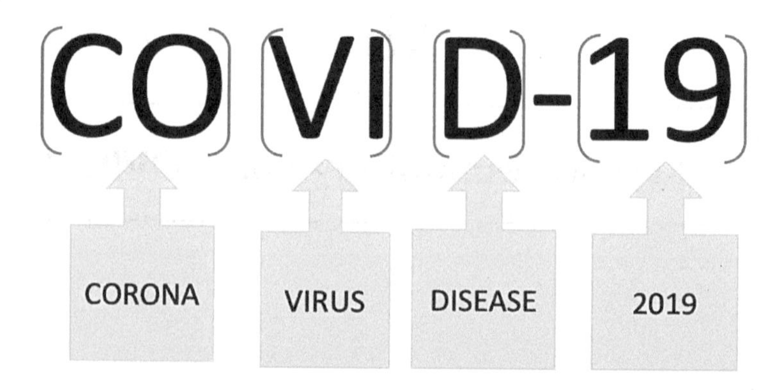

Figure 15.1: COVID-19 Naming

A blockchain is an essential tool for establishing an efficient and transparent healthcare business model based on higher degrees of accuracy and trust because technology is a tamper-proof public ledger. Blockchain will surely not prevent the emergence of new viruses itself, but what it can do is create the first line of rapid protection through a network of connected devices whose primary goal is to remain alert about disease outbreaks. Therefore, the use of blockchain-enabled platforms can help prevent these pandemics by enabling early detection of epidemics, fast-tracking drug trials, and impact management of outbreaks and treatment. [72]

But before we explore in details the possible ways of using Blockchain to help in fighting this invisible enemy, we need to understand some of the challenges defining this deadly virus.

15.1 Major Challenges of COVID-19

- One major issue is how prepared the world's health systems are to respond to this outbreak.

- Tracking a huge population of infectious patients to stop epidemics.
- Another is the immediate requirement for developing better diagnostics, vaccines, and targeted therapeutics.
- Misinformation and conspiracy theories spread through social media platforms.
- Various limitations while accessing the tools when required.
- No adequate measures to adopt in a crisis situation. [8][73][74]

15.2 Can Blockchain help in preventing pandemics?

With Blockchain we can share any transaction / information, real time, between relevant parties present as nodes in the chain, in a secure and immutable fashion. In this case, had there been a blockchain where WHO, Health Ministry of each country and may be even relevant nodal hospitals of each country, were connected, sharing real time information, about any new communicable disease, then the world might have woken up much earlier. We might have seen travel restrictions given sooner, quarantining policies set sooner and social distancing implemented faster. And may be fewer countries would have got impacted.

What every country is doing now fighting this pandemic, would have been restricted to fewer countries and in a much smaller scale. The usage of a Blockchain to share the information early on, might have saved the world a lot of pain. [75]

The world had not seen anything like COVID-19 pandemic before in the recent history. Today we need to take a hard look at the reporting infrastructure available for communicable diseases, both technology and regulations and improve upon that, such that we do not need to face another pandemic like this in the future. Figure 15.2 list the areas where Blockchain Technology can help in fighting COVID19. [68]

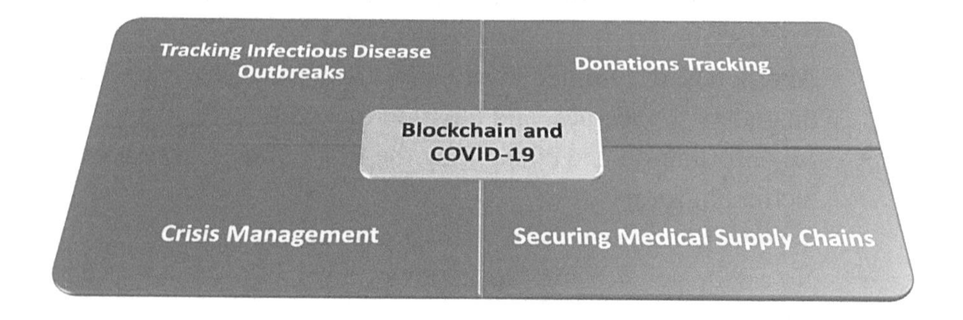

Figure 15.2: Blockchain Applications in fighting COVID-19

15.3 Tracking Infectious Disease Outbreaks

Blockchain can be used for tracking public health data surveillance, particularly for infectious disease outbreaks such as COVID-19. With increased blockchain transparency, it will result in more accurate reporting and efficient responses. Blockchain can help develop treatments swiftly as they would allow for rapid processing of data, thus enabling early detection of symptoms before they spread to the level of epidemics.

Additionally, this will enable government agencies to keep track of the virus activity, of patients, suspected new cases, and more. [76] [68]

15.4 Donations Tracking

As trust is one of the major issues in donations, Blockchain has a solution for this issue.

There has been a concern that the millions of dollars being donated for the public are not being put to use where needed.

With the help of blockchain capabilities, donors can see where funds are most urgently required and can track their donations until they are provided with a verification that their contributions have been received to the victims. Blockchain would enable transparency for the general public to understand how their donations have been used and its progress. [68] [70]

15.5 Crisis Management

Blockchain could also manage crisis situation. It could instantly alert the public about the Coronavirus by global institutes like the World Health Organization (WHO) using smart contracts concept.

Not only it can alert, but Blockchain could also enable to provide governments with recommendations about how to contain the virus. It could offer a secure platform where all the concerning authorities such as governments, medical professionals, media, health organizations, media, and others can update each other about the situation and prevent it from worsening further. [77] [68]

15.6 Securing Medical Supply Chains

Blockchain has already proven its success stories as a supply chain management tool in various industries; similarly, Blockchain could also be beneficial in tracking and tracing medical supply chains.

Blockchain-based platforms can be useful in reviewing, recording, and tracking of demand, supplies, and logistics of epidemic prevention materials. As supply chains involve multiple parties, the entire process of record and verification is tamper-proof by every party, while also allowing anyone to track the process.

This technology could help streamline medical supply-chains, ensuring that doctors and patients have access to the tools whenever they need them, and restraining contaminated items from reaching stores.[68] [72]

15.7 WHO and Blockchain Technology

The World Health Organization (WHO) is working with blockchain and other tech companies on a program to help convey data about the ongoing COVID-19 pandemic, named MiPasa.

The program is a distributed ledger technology (DLT) that will hopefully help with early detection of the virus and identifying carriers and hotspots.

MiPasa is built on top of Hyperledger Fabric in partnership with IBM, computer firm Oracle, enterprise blockchain platform HACERA and IT corporation Microsoft. It purports to be "fully private" and share information between need-to-know organizations like state authorities and health officials.

Described by creators as "an information highway," MiPasa cross-references siloed location data with health information. It promises to protect patient privacy and to help monitor local and global trends such as the virus that has now sent the world spiraling into chaos and uncertainty in recent weeks.

The U.S., European, and Chinese Centers for Disease Control and Prevention, the Hong Kong Department of Health, the Government of Canada and China's National Health Commission have all worked with the project. [8][68] [69][71]

References

[01] https://en.wikipedia.org/wiki/Gossip_protocol

[02] https://medium.com/the-daily-bit/9-types-of-consensus-mechanisms-that-you-didnt-know-about-49ec365179da

[03] https://www.businessinsider.com/blockchain-technology-applications-use-cases-2017-9

[04] https://www.disruptordaily.com/blockchain-use-cases-crowdfunding/

[05] https://due.com/blog/a-new-era-of-crowdfunding-blockchain/

[06] https://www.forbes.com/sites/bernardmarr/2018/03/23/how-Blockchain-will-transform-the-supply-chain-and-logistics-industry/#c-7c357e5fecd

[07] https://blockgeeks.com/guides/Blockchain-and-supply-chain/

[08] https://www.technologyreview.com/2017/01/05/5880/a-secure-model-of-iot-with-blockchain/

[09] https://consensys.net/Blockchain-use-cases/supply-chain-management/

[10] https://en.wikipedia.org/wiki/Bitcoin_Cash

[11] https://www.bitdegree.org/tutorials/bitcoin-cash-vs-bitcoin/

[12] https://btcmanager.com/us-authorities-blockchain-covid-19-critical-services/?q=/us-authorities-blockchain-covid-19-critical-services/&

[13] https://www.investopedia.com/terms/s/smart-contracts.asp

[14] https://hackernoon.com/everything-you-need-to-know-about-smart-contracts-a-beginners-guide-c13cc138378a

[15] https://solidity.readthedocs.io/en/v0.4.24/introduction-to-smart-contracts.html

[16] https://www.gartner.com/newsroom/id/3123018

[17] https://www.amazon.com/Secure-Smart-Internet-Things-IoT/dp/8770220301/

[18] https://techcrunch.com/2016/06/28/decentralizing-iot-networks-through-blockchain/

[19] http://www-935.ibm.com/services/multimedia/GBE03662USEN.pdf

[20] https://www.accenture.com/us-en/new-applied-now

[21] https://www.computerworld.com/article/3027522/internet-of-things/beyond-bitcoin-can-the-blockchain-power-industrial-iot.html

[22] http://www.treasuryandrisk.com/2017/03/09/blockchain-technology-balancing-benefits-and-evolv?slreturn=1507334668&page=4

[23] http://www.livebitcoinnews.com/three-risks-assess-company-considering-blockchain/

[24] https://hbr.org/2017/03/how-safe-are-blockchains-it-depends

[25] https://www.i-scoop.eu/digital-transformation/

[26] https://www.bbc.com/news/business-16611040

[27] https://pdfs.semanticscholar.org/presentation/cc44/7518775fac6a1c1b173631c4200adb451d39.pdf

[28] http://www.zdnet.com/article/why-ai-and-machine-learning-need-to-be-part-of-your-digital-transformation-plans/

[29] https://iot.ieee.org/newsletter/january-2017/iot-and-blockchain-convergence-benefits-and-challenges.html

[30] https://datafloq.com/read/how-blockchain-secure-model-internet-of-things/2583

[31] https://www.allerin.com/blog/4-myths-associated-with-blockchain

[32] https://home.kpmg.com/uk/en/home/insights/2017/04/five-blockchain-myths-that-just-wont-die.html

[33] http://www.techrepublic.com/article/five-big-myths-about-the-bitcoin-blockchain/

[34] https://www.forbes.com/sites/yec/2017/05/04/debunking-blockchain-myths-and-how-they-will-impact-the-future-of-business/#275f5ef05609

[35] http://www.reuters.com/article/us-blockchains-technology-commentary-idUSKCN0Y22GC

[36] https://www.wired.com/story/cyberinsurance-tackles-the-wildly-unpredictable-world-of-hacks/

[37] https://www.whitehouse.gov/wp-content/uploads/2018/03/The-Cost-of-Malicious-Cyber-Activity-to-the-U.S.-Economy.pdf

[38] https://www.infosecurity-magazine.com/next-gen-infosec/blockchain-technology/

[39] https://www.allerin.com/blog/blockchain-enabled-smart-contracts-all-you-need-to-know

[40] https://www.ibm.com/blogs/insights-on-business/government/convergence-blockchain-cybersecurity/

[41] https://www.forbes.com/sites/rogeraitken/2017/11/13/new-blockchain-platforms-emerge-to-fight-cybercrime-secure-the-future/#25bdc5468adc

[42] https://blog.capterra.com/benefits-of-blockchain-cybersecurity/

[43] http://www.technologyrecord.com/Article/cybersecurity-via-blockchain-the-pros-and-cons-62035

[44] https://www.bbntimes.com/en/technology/first-line-of-defense-for-cybersecurity-artificial-intelligence

[45] https://blog.goodaudience.com/blockchain-and-artificial-intelligence-the-benefits-of-the-decentralized-ai-60b91d75917b

[46] https://aibusiness.com/ai-brain-iot-body/

[47] https://www.forbes.com/sites/darrynpollock/2018/11/30/the-fourth-industrial-revolution-built-on-blockchain-and-advanced-with-ai/#4cb2e5d24242

[48] https://www.forbes.com/sites/rachelwolfson/2018/11/20/diversifying-data-with-artificial-intelligence-and-blockchain-technology/#1572eefd4dad

[49] https://www.coindesk.com/how-should-crypto-prepare-for-googles-quantum-supremacy?

[50] https://ai.googleblog.com/2019/10/quantum-supremacy-using-programmable.html

[51] https://www.linkedin.com/pulse/20140503185010-246665791-quantum-computing/

[52] https://decrypt.co/9642/what-google-quantum-computer-means-for-bitcoin/

[53] https://www.etorox.com/news/opinions/the-7-key-takeaways-about-libra/

[54] https://www.coindesk.com/theres-a-second-token-a-breakdown-of-facebooks-blockchain-economy

[55] https://www.cnbc.com/2019/06/18/what-is-libra-facebooks-new-cryptocurrency.html?__source=facebook%7Cmakeit

[56] https://techcrunch.com/2019/06/18/facebook-libra/?tpcc=ECFB2019

[57] https://insurtechforum.net/compared-facebook-libra-and-calibra-vs-apple-pay/?

[58] https://www.bloomberg.com/news/articles/2019-06-25/facebook-s-libra-creates-new-path-to-printing-even-more-money-through-messaging-services?

[59] https://www.nytimes.com/2019/06/19/opinion/facebook-currency-libra.html?smid=tw-nytopinion&smtyp=cur

[60] https://www.forbes.com/sites/jonathanberr/2019/06/30/will-facebooks-new-cryptocurrency-enable-foreign-election-meddling/#19042b237949

[61] https://www.coindesk.com/billion-dollar-returns-the-upside-of-facebooks-libra-cryptocurrency

[62] https://www.bloomberg.com/news/articles/2019-06-23/guardians-of-money-bristle-at-zuckerberg-s-new-financial-order?

[63] https://www.theatlantic.com/ideas/archive/2019/06/dont-trust-libra-facebooks-new-cryptocurrency/592450/

[64] https://aithority.com/guest-authors/Blockchain-technology-in-the-future-7-predictions-for-2020/

[65] https://www.forbes.com/sites/andrewarnold/2018/08/13/the-6-major-Blockchain-trends-for-2018-outlined-by-deloitte/#40ac48474844

[66] https://www.cio.com/article/3294225/Blockchain/5-top-Blockchain-trends-of-2018.html

[67] https://www.bbntimes.com/en/technology/second-line-of-defense-for-cybersecurity-blockchain

[68] https://www.forbes.com/sites/forbestechcouncil/2018/07/18/three-breakthroughs-that-will-disrupt-the-tech-world-in-2019/#485c59e61f87

[69] https://achievion.com/blog/5-trends-Blockchain-application-development-2018.html

[70] https://hackernoon.com/latest-Blockchain-trends-to-watch-out-for-in-2018-59b2831ddfb

[71] https://www.govtech.com/products/Blockchain-Emerges-as-Useful-Tool-in-Fight-Against-Coronavirus.html

[72] https://www.blockchain-council.org/blockchain/how-blockchain-can-solve-major-challenges-of-covid-19-faced-by-healthcare-sectors/

[73] https://www.ibm.com/blogs/blockchain/2020/03/mipasa-project-and-ibm-blockchain-team-on-open-data-platform-to-support-covid-19-response/

[74] https://www.pymnts.com/blockchain/bitcoin/2020/bitcoin-daily-who-debuts-mipasa-blockchain-to-share-covid-19-data-coinbases-retail-payments-portal-passes-200m-transactions-processed/

[75] https://mipasa.org/about/

[76] https://www.ledgerinsights.com/us-homeland-security-lists-blockchain-as-covid-19-critical-service/

[77] https://www.rollcall.com/2020/03/31/blockchain-could-transform-supply-chains-aid-in-covid-19-fight/

Index

D

E

F

G

H

I

About the Author

Prof. Ahmed Banafa

IoT-Blockchain –AI- Cybersecurity Expert | Award-Winning Author | Faculty | Keynote Speaker

Prof. Ahmed Banafa has extensive experience in research, operations, and management, with a focus on IoT, Blockchain, Cybersecurity, and AI. His researches cited in studies by international organizations including: NATO, WTO, and APEC. He is a reviewer and a technical contributor for several technical books. He served as an instructor at well-known universities and colleges, including Stanford University, University of California, Berkeley; California State University-East Bay; San Jose State University; and the University of Massachusetts. He is the recipient of several awards, including Distinguished Tenured Staff Award, Instructor of the year for 4 years in a row, and Certificate of Honor from the City and County of San Francisco. He was named as No.1 tech voice to follow, technology fortune teller and influencer by LinkedIn in 2018 by LinkedIn, his research featured in many reputable sites and magazines including Forbes, IEEE, and MIT Technology Review, and Interviewed by CNN, ABC, CBS, NBC, BBC, NPR and Fox TV and Radio stations. He is a member of the MIT Technology Review Global Panel. He studied Electrical Engineering at Lehigh University, Cybersecurity at Harvard University and Digital Transformation at Massachusetts Institute of Technology (MIT). He is the author of the books: *Secure and Smart Internet of Things (IoT) using Blockchain and Artificial Intelligence (AI)*, and *Blockchain Technology and Applications.* Winner of *Author & Artist Award 2019* of San Jose State University for "Secure and Smart IoT" Book.